CURRENT AFFAIRS:
PERSPECTIVES ON ELECTRICITY POLICY FOR ONTARIO

Edited by Doug Reeve, Donald N. Dewees, and Bryan W. Karney

In Ontario, electricity has traditionally been generated from hydroelectric facilities, coal, and nuclear power. Today the province must respond to the escalating demand for electricity, coupled with an aging infrastructure and diminishing capacity, while also addressing growing concerns about air pollution, global warming, and the environmental impact of fossil fuel combustion. Sources of hydroelectric power in the province are limited, while nuclear power is expensive and raises safety concerns. Power from green energy sources, such as the sun and wind, is also costly to produce and in some cases is intermittent. Electricity policy in Ontario thus requires the careful balancing of environmental goals and values against costs that must be borne by consumers and taxpayers.

Current Affairs brings together local and international experts on energy, the environment, and public policy to discuss the major electricity policy issues facing Canada's most populous and industrial province. The contributors, who represent academia, think tanks, consulting firms, governments, and the International Energy Agency, bring fresh perspectives and a wealth of experience to the substantive questions concerning Ontario's electricity future. Collectively they suggest that in Ontario, as in other jurisdictions, environmental policy must be connected to meaningful research, public dialogue, and solid educational and regulatory programs to support the behavioural and institutional changes that will lead to a sustainable electricity future.

DOUG REEVE is a professor and chair of the Department of Chemical Engineering and Applied Chemistry at the University of Toronto.

DONALD N. DEWEES is a professor in the Department of Economics at the University of Toronto.

BRYAN W. KARNEY is chair of the Division of Environmental Engineering and Energy Systems and a professor in the Department of Civil Engineering at the University of Toronto.

CURRENT AFFAIRS

Perspectives on Electricity Policy for Ontario

Edited by Doug Reeve, Donald N. Dewees, and Bryan W. Karney

UNIVERSITY OF TORONTO PRESS
Toronto Buffalo London

Toronto Buffalo London
www.utppublishing.com
Printed in Canada

ISBN 978-1-4426-4019-1 (cloth)
ISBN 978-1-4426-0994-5 (paper)

Printed on acid-free, 100% post-consumer recycled paper with vegetable-based inks.

Library and Archives Canada Cataloguing in Publication

Current affairs : perspectives on electricity policy for Ontario / edited by Doug Reeve, Donald N. Dewees, and Bryan W. Karney.

Includes bibliographical references.
ISBN 978-1-4426-4019-1 (bound). – ISBN 978-1-4426-0994-5 (pbk.)

1. Electric power – Ontario. 2. Electric utilities – Government policy – Ontario. 3. Electric power – Environmental aspects – Ontario. I. Reeve, Doug II. Dewees, Donald N., 1941– III. Karney, Bryan William, 1957–

HD9685.C3C87 2010 333.793′21509713 C2009-906820-6

University of Toronto Press acknowledges the financial assistance of the Canada Council for the Arts and the Ontario Arts Council to its publishing program.

University of Toronto Press acknowledges the financial support of its publishing activities by the Government of Canada through the Book Publishing Industry Development Program (BPIDP).

Contents

Part Three: Finding the Right Price

Part Four: Policy Tools for Increasing End-Use Electricity Efficiency

Part Five: Inter-jurisdictional Cooperation in Achieving Energy Policy Goals

Preface

DOUG REEVE AND DONALD N. DEWEES

In early 2007, at the University of Toronto, a new school of public policy and governance was being established, and the Faculty of Applied Science and Engineering was exploring the prospects of engineering and public policy. A conversation started about an opportunity to draw upon the university in a new way: to have the academy apply its capability for research, analysis, debate, and discussion to electricity policy broadly, but also in a more focused way applied to Ontario. It was suggested that the School of Public Policy and Governance might create a forum that would generate a fresh perspective by bringing new people to the Ontario electricity policy discussion. The school brought together a team from the University of Toronto (from engineering, economics, law, political science, environmental studies, and management), from the electricity agencies, and from the government to frame the questions for debate, to develop a program, and to identify potential contributors. It was a priority to bring to the discussion relevant and useful experiences from other jurisdictions that could then be focused on the Ontario scene. The team met many times to prepare the workshop 'Current Affairs: Perspectives on Electricity Policy for Ontario,' held on 4–5 June 2008. The workshop provided the core content for this book.

The workshop was organized to bring fresh information and insight but also to stimulate debate and discussion. In each of the four main sessions a session chair framed the questions, a main speaker or two (in all cases from beyond Ontario's boundaries) addressed the questions from the perspective of his or her experience, and a discussant or two brought the speaker's remarks into the context of Ontario. A keynote talk during the evening before the workshop set the stage, and summary remarks on Ontario electricity policy and politics closed the workshop. The program and presentations are archived on the Internet at http://www.publicpolicy.utoronto.ca/Electricity/index.html.

Contributors to the Discussion

The principal workshop speakers were drawn from Europe and the United States and brought with them invaluable experience from France, the United Kingdom, and the Nordic countries, and across the United States. They brought a wealth of expertise from the academy, think tanks, consulting firms, governments, and the International Energy Agency (IEA). The following were the principal workshop speakers, in order of speaking:

- Ted Parson, a professor at the Law School and the School of Natural Resources and Environment of the University of Michigan (who delivered the keynote address); he has been engaged in international policymaking on behalf of the governments of Canada and the United States.
- Carolyn Fischer, from Resources for the Future, an independent research group studying environmental and natural resource policy, in Washington, DC; she has worked on modelling and economic evaluation of environmental policies.
- Dominique Finon, Deputy Director of the energy program at the French National Centre of Scientific Research (CNRS); he has acted as an expert for the European Commission and is particularly interested in policies for renewable electricity generation, energy efficiency, and market reforms.
- Branko Terzic, Regulatory Policy Leader, Energy and Resources, Deloitte Services in Washington, DC; he is a former commissioner of the U.S. Federal Energy Regulatory Commission.
- Ulrik Stridbaek, who has worked as a senior policy advisor, electricity markets, at the International Energy Agency in Paris. His most recent publications are the IEA books *Lessons from Liberalised Electricity Markets* (2005) and *Tackling the Investment Challenges in Power Generation* (2007).
- Loren Lutzenhiser, Professor of Urban Studies and Planning at Portland State University; he recently completed a major study for the California Energy Commission, reporting on the behaviour of households, businesses, and governments in the aftermath of that state's 2001 electricity deregulation crisis.

- Steve Sorrell, a senior fellow at the University of Sussex; he is currently leading the governance theme of the Sussex Energy Group and co-managing the technology and policy assessment function of the UK Energy Research Centre.
- Perry Sioshansi, from the consulting firm Menlo Energy Economics, which serves utility clients and policymakers out of San Francisco; he is a specialist in electric power sector restructuring.

The following discussants brought their knowledge of Ontario issues and provided focused reflection to the principal speakers' talks (in order of presentation):

- James Meadowcroft, Professor and Canada Research Chair in Governance for Sustainable Development at Carleton University.
- Tom Adams, a consultant, formerly executive director of Energy Probe.
- Mark Jaccard, Professor of Environmental Economics at the School of Resource and Environmental Management at Simon Fraser University.
- Dean Mountain, Professor and Director of the McMaster Institute for Energy Studies at McMaster University.
- George Vegh, a lawyer with McCarthy Tétrault's Toronto energy regulatory practice and formerly general counsel for the Ontario Energy Board.

The workshop was closed by two eminent figures in Ontario electricity circles:

- Sean Conway, former Ontario MPP and Cabinet minister, now with Queen's University School of Policy Studies.
- Michael Trebilcock, a professor in law and economics and University Professor at the University of Toronto.

We note that several environmental organizations have made important, thoughtful contributions to the electricity policy debate in Ontario, in particular Ontario Clean Air Alliance, Pembina Institute, and WWF Canada, and we regret that, other than discussant Tom Adams, non-governmental organization researchers did not have a greater role in this work. We also recognize that although several of the discussants and concluding speakers were from Ontario universities, there are numerous other academics in Ontario making important contributions to electricity policy. Recognizing the need for more information exchange and the inclusion of a larger number and wider range of policy researchers, we have taken steps, with James Meadowcroft and others, to create a new forum, the Ontario Network for Sustainable Energy Policy.

The Book Takes Shape

The purpose of *Current Affairs* is to bring fresh perspectives to the substantive questions concerning Ontario's electricity future; to articulate the environmental, economic, and political challenges; and to examine innovative and pragmatic policy options.

The book is organized around the four sessions of the workshop: part 2, 'Electricity's Role in Reducing the Environmental Footprint of Energy Use'; part 3, 'Finding the Right Price'; part 4, 'Policy Tools for Increasing End-use Electricity Efficiency'; and part 5, 'Inter-jurisdictional Cooperation in Achieving Energy Policy Goals.' In each part there is an introduction by the session chair, a chapter by each main speaker, and chapters by the discussant(s). We note with regret that the text of part 4 is more modest than intended owing to missing text versions of the presentations by Tom Adams and Branko Terzic. The opening and closing parts of the book, respectively, set the stage and draw conclusions: part 1, 'A Global Perspective on Electricity Policy for Ontario,' and part 6, 'Policy Challenges and Opportunities.'

Current Affairs is intended to enhance the discussion of electricity policy issues among all those interested in the successful evolution of the electricity sector in Ontario. It should be of interest to all those involved in electricity and electricity policy: politicians and policy-makers, academics and researchers, consultants, lawyers, bankers, aboriginal peoples, consumers big and small, and those in industry, utilities, unions, and not-for-profit organizations.

Abbreviations

ACEEE	American Council for an Energy-Efficient Economy
AMI	advanced metering infrastructure
BIEE	British Institute of Energy Economics
CAFE	Corporate Average Fuel Economy
CAISO	California's independent system operator
CCGT	combined cycle gas technology
CCS	carbon capture and sequestration/storage
CEC	California Energy Commission
CFL	compact fluorescent lamp
CGE	computable general equilibrium
CHP	combined heat and power
CIEE	University of California Institute for Energy and the Environment
CMI	The Cambridge and MIT Institute
CO_2	carbon dioxide
DOE	U.S. Department of Energy
DR	demand response
DSM	demand side management
EC	European Commission
EC/DG TREN	European Commission Directorate-General of Transport and Energy
ECEEE	European Council for an Energy Efficient Economy
ECX	European Climate Exchange
EDF	Electricité de France
EE	energy efficiency
EJ	exajoules

E.ON	Elektrizität.on (largest electricity and gas company in Germany)
EPR	European pressurized water reactor
EPRI	Electric Power Research Institute
ETS	Emission Trading System
EU	European Union
EU-15	European Union with the first fifteen member states (before 2001)
EWEA	European Wind Energy Association
FERC	U.S. Federal Energy Regulatory Commission
FIT	feed-in tariff
GAM	Global Adjustment Mechanism
Gw	gigawatt
HOEP	hourly Ontario electricity price
IEA/NEA	International Energy Agency / Nuclear Energy Agency
IEA-OECD	International Energy Agency – Organisation for Economic Cooperation and Development
IESO	Independent Electricity System Operator
IMO	Independent Market Operator
IPCC	Intergovernmental Panel on Climate Change
IPSP	Integrated Power System Plan
kt	kiloton
kW	kilowatt
kWh	kilowatt-hour
LCA	life-cycle analysis/assessment
LCFS	Low Carbon Fuel Standard
MIT	Massachusetts Institute of Technology
MRTS	marginal rate of technical substitution
MT	market transformation
MW	megawatt
MWh	megawatt-hour
NAP	national allocation plans
NEA	*National Energy Act*
NEEA	Northwest Energy Efficiency Alliance
NPCC	Northwest Power and Conservation Council
NRCAN	Natural Resources Canada
OEB	Ontario Energy Board
OECD	Organisation for Economic Co-operation and Development

OPA	Ontario Power Authority
OPG	Ontario Power Generation
OPTRES Project	'"OPTimisation of REnewable Support schemes" in the European electricity market' of the European Commission, 2004–6
PHEV	plug-in hybrid electric vehicle
PJM	Pennsylvania-New Jersey-Maryland electricity market pool
PNNL	Pacific Northwest National Laboratory
ppm	parts per million
PTEM	physical-technical-economic model
RD&D	research, development, and demonstration
RES	renewable energy sources
RES-E	renewable energy sources in electricity
RPS	renewable portfolio standard
RWE	Rheinisch-Westfalisches Elektrizitätswerk
TGC	tradable green certificate
TSO	transmission system operator
TWh	terawatt-hour
VAT	value added tax
W	watt

CURRENT AFFAIRS:
PERSPECTIVES ON ELECTRICITY POLICY FOR ONTARIO

1 The Evolution of Ontario Electricity Policy

DONALD N. DEWEES AND DOUG REEVE

If we were not concerned about air pollution and global warming, Ontario's electricity policy would be simple – burn coal. Coal power costs less than other available new electricity sources, and both Canada and the United States have abundant supplies of coal. However, we are concerned about conventional air pollutants – particulate matter, sulphur dioxide, nitrogen oxides, and toxic metals such as cadmium and mercury that harm human health and damage the environment. We are increasingly concerned about global warming, and coal-fired power plants discharge large amounts of carbon dioxide. Vigorous pursuit of environmental protection calls into question the continuation of fossil fuel combustion for electricity generation in Ontario. Technology being developed may permit further reductions of conventional pollution emissions from coal-fired power plants, but as sequestration of carbon dioxide remains problematic in Ontario, there is not yet an obvious 'clean coal' solution to Ontario's electricity demand. The cost of green power from wind or sun is very high, and it is not clear how quickly or how much these costs will decline. Considerable hydroelectric power is available in Manitoba and Quebec; however, the great distance to Manitoba's power means that it would be costly to bring it to major Ontario users. Quebec can export its hydropower to New York and New England at a price that is substantially higher than the Ontario price. Small hydroelectric power is expensive. Nuclear power raises its own environmental and safety concerns, and Ontario's experience with nuclear power has been expensive. As a result, electricity policy requires the careful balancing of environmental goals and values against costs that must be borne by consumers and taxpayers. Ontario's electricity policy is unfortunately not simple; it is complex and contentious.

Today's policy challenges are vastly different from those of much of the twentieth century. Low-cost electricity was central to the economic development of the province until late in the twentieth century as Ontario exploited inexpensive power from the hydroelectric facilities at Niagara Falls and from other major rivers. Cheap power attracted energy-intensive industry. However, by the 1960s the demand was growing beyond the capacity of Ontario's low-cost hydro facilities, so we built fossil and nuclear plants that produced power no less expensively than that produced in neighbouring states. Today the population and the economy of the province continue to grow, particularly in the Toronto region, but meeting increased electricity demand requires increasing the development of higher-cost sources. Higher costs and environmental concerns, including global warming and opposition to new generating facilities and new transmission lines, have led to a desire to use electricity more efficiently than we have in the past and thus to invest in conservation strategies. The province's electricity infrastructure is undergoing fundamental overhaul to deal with these and other problems such as aging nuclear plant reliability and cost, coal plant emissions, aging and load-limited transmission lines, and a desire to expand renewable power.

In 2009, Ontario policy includes a commitment to close the large Nanticoke coal-fired generating station by 2014 and to be 90% carbon free by 2027. The *Green Energy and Green Economy Act* (Bill 150) has just been introduced at the time of writing and, when passed into law, promises to invigorate and expedite renewable generation. We believe that the commitments to decreasing carbon dioxide output will be hard to achieve. It will be difficult to provide capacity, and reduce demand sufficiently to allow that generation station closure, at a price Ontarians want to pay. The coal closure has been postponed twice already, and some observers expect that it will be postponed again as we fail to meet conservation targets and fail to commission sufficient alternative generation facilities. The generation mix that would achieve the 90% carbon-free goal by 2027 is controversial and poses substantial risks of both cost and feasibility. Reducing fossil fuel consumption for transportation, space heating, and industrial process heating in order to decrease carbon dioxide output is likely to increase the demand for electricity (for example, by electric cars), thereby increasing the stress on the electricity sector. So, the existence of green commitments for electricity does not dispose of the need to craft more detailed policies that will either implement those plans or modify them if needed. Those who support these goals worry about how we will meet them. Canada has adopted a series of federal plans to reduce greenhouse gas emissions in pursuit of our Kyoto commitment, yet we have failed to meet that commitment and, indeed, have failed to prevent our greenhouse gas emissions from increasing steadily. This failure deepens our concerns that Ontario might fall short of its commitments.

Current Affairs and the conference that gave rise to it are about how to develop electricity policy to meet these challenging environmental goals. Most sections of this volume deal directly or indirectly with the intersection of environmental policy and electricity policy. Part 5 deals with inter-jurisdictional relations, because Ontario borders on five states and two provinces and we exchange electricity and air pollution with our neighbours. Ontario policy cannot be developed without careful consideration of what our neighbours are doing with respect to both environment and electricity and how our policies interact. Several European contributors to this volume share experience that is relevant to Ontario.

Throughout the book there are references to the cost, performance, or environmental characteristics of various electricity-generation technologies. Our goal, however, is not to pick the 'right' technology or mixture of technologies. There are disputes about the cost and performance of various technologies today. Predicting future cost and performance will be more difficult. We believe that identification of the right technology should evolve over time through the implementation of effective policies. This volume is about how to design and implement effective policies that will in the long run lead to adoption of the most appropriate technologies for Ontario.

Policy Background

An extensive review of the history of electricity policy in Ontario is beyond the scope of this volume, but a brief review of electricity restructuring and of environmental regulation during the last two decades may help to explain why we are in our present situation and may inform our sense of what is feasible Ontario policy.[1] The *Power Corporation Act* required Ontario Hydro to provide 'power at cost.' The Government of Ontario appointed the directors of Ontario Hydro and guaranteed its debt and could issue policy directives to Ontario Hydro. Ontario Hydro had long been close to the government. The construction of large nuclear plants and high-voltage lines in the Greater

Toronto Area had raised concerns about electric power planning, and Dr Omond Solant was asked to review capital planning at Ontario Hydro (Ontario Power Authority 2005, 4). The 1974 Solant Commission report recommended that public consultation become a significant part of the electricity planning process. In 1980 the Porter Commission report on supply planning recommended that formal long-range plans be developed and that they include demand management as well as capacity expansion. In 1989 Ontario Hydro published a 25-year demand-supply plan considering a wide range of generation options to meet rapidly growing future demand. By 1992 the plan had to be revised because of a fall in forecast demand. The high cost of Darlington Nuclear Generation Station, combined with below-forecast demand, drove up electricity prices by 30% in the early 1990s, and the resulting public outcry led the government to freeze the price from 1993 until 2002, causing deficits for Ontario Hydro. Concern grew that Ontario Hydro was no longer performing as well as it should.

In 1995, the Government of Ontario appointed a committee to study options for competition in Ontario's electricity system (ACCOES 1996). The government issued a white paper and then appointed the Market Design Committee, which recommended an initial market design and a set of market rules (Ontario Market Design Committee 1999). During the 1990s, nuclear units experienced operating problems, and by 1998 eight nuclear units at Pickering A and Bruce A generating stations were out of service for serious maintenance and upgrading. For years there was debate as to the economic merits of refurbishing or closing down each of these units while their incapacity caused supply constraints in Ontario and the cost risks overhung electricity policy discussions. In anticipation of a competitive market, legislation in 1998 divided Ontario Hydro into a transmission company (Hydro One), a generation company (Ontario Power Generation [OPG]), and an Independent Market Operator (IMO). Competitive wholesale and retail markets were developed. It was expected that markets would replace central planning of the electricity system.

The government opened Ontario's competitive wholesale market on 1 May 2002, passing the spot price to any consumer who had not signed up with a retailer for a fixed price. Record drought and heat waves in the summer of 2002 combined with a shortage of nuclear units (the eight units were not back in service) to drive the peak hourly price to $4.71 per kilowatt-hour (kWh) on 2 July, more than one hundred times the normal price (Independent Electricity Market Operator

2002, 5). Monthly average prices nearly tripled from May to September. Consumers were outraged, and on 11 November 2002 the government froze the price for smaller consumers at $0.043/kWh. The price freeze was extended to somewhat larger consumers in 2003, and the price was raised gradually over several years. The price that OPG would receive for the output of its nuclear units and its baseload hydroelectric units was regulated by the Ontario Energy Board (OEB).

The supply shortage led to the appointment of the Electricity Conservation and Supply Task Force in 2003, which predicted, in 2004, a serious supply shortfall. The government responded by contracting for new electricity supply. Legislation in 2004 created the Ontario Power Authority with a mandate to deal with deficiencies in the competitive market, including long-term planning and energy conservation. It has produced the *Integrated Power System Plan* (IPSP) and has contracted for new supply, as well as initiating conservation and demand management programs (Ontario Power Authority 2005, 9). The IPSP is a twenty-year plan that specifically identifies the supply mix from conventional sources, conservation, demand side management, and renewable resources. The result by 2008, discussed by George Vegh in chapter 10, is a hybrid structure in which a spot market sets part of the price, the OEB regulates some prices, a crown corporation (OPG), dominates generation, and the OPA manages a competitive bidding process for several types of new power supply.

Electricity generation in Ontario has long been subject to regulations limiting air pollution emissions including general regulations applied to all stationary sources. In 1986, Ontario Hydro, along with three other major corporate sources of acid gas emissions, became subject to the Countdown Acid Rain program, which limited its sulphur dioxide emissions to 370 kilotons (kt) per year in 1986, 240 kt per year in 1990, and 175 kt per year in 1994 and subsequent years. Nitric oxide emissions were also limited.[2] In 1995 Ontario Hydro made a voluntary commitment to reduce its greenhouse gas emissions to its 1990 levels by the year 2000, and it began working to reduce those emissions. By the late 1990s the Ontario Clean Air Alliance was lobbying vigorously and effectively for an end to coal combustion for electricity generation, rather than for the regulation of the emission rate of individual pollutants. In 2000 OPG began purchasing and retiring carbon dioxide emission credits to meet its voluntary target (Ontario Power Generation 2001). In 2001 the Ontario Ministry of the Environment adopted an emission trading system for sulphur dioxide and nitrogen oxides for the electricity sector and in 2005 expanded it to include major industrial sectors.[3] In 2003 Premier Dalton McGuinty made a commitment that Ontario would be off coal for electricity generation by 2007. The motive at the time was avoiding the health consequences of traditional air pollutants. Less than two years later, the phase-out of coal was postponed to 2009, and the motivation was expanded to include reductions in greenhouse gas emissions. In 2005 the coal-fired Lakeview Generating Station was closed and its four smokestacks demolished, dramatically symbolizing the permanence of the closure. The closing of Nanticoke Generating Station was postponed to 2014 as the difficulty of earlier closure became apparent (Ontario Ministry of Energy and Infrastructure 2009).

Exploring Electricity Policy

The concerns articulated above have led us to a series of questions:

- How do we reduce the environmental footprint of the electricity system?
- What is the right price for electricity?
- How can we increase end-use electricity efficiency?
- What is the best way to integrate electricity supply (and environmental policies) across jurisdictional boundaries?

These important questions and the challenging policy situation in Ontario are addressed in the following chapters.

NOTES

1 For a more extensive history of electricity institutions and policy in Ontario, see ACCOES (1996), Daniels and Trebilcock (1996), Trebilcock and Hrab (2005), Dewees (2009), Doern (2005), OPA (2005) , and Wikipedia (2009).

2 See Ontario Regulation 281/87.

3 See *Ontario Regulations 397/01 and 194/05. See Homsey 2008 for an analysis of the history of emissions trading in Ontario.*

REFERENCES

ACCOES (Advisory Committee on Competition in Ontario's Electricity System). 1996. *A framework for competition.* Toronto: Queen's Printer.

Daniels, Ronald J., and Michael J. Trebilcock. 1996. 'The future of Ontario Hydro: A review of structural and regulatory options.' In *Ontario Hydro at the millennium: Has monopoly's moment passed?* ed. Ronald J. Daniels. Kingston: McGill-Queen's University Press.

Dewees, Donald N. 2009. 'Electricity restructuring in the provinces: Pricing, politics, starting points, and neighbours.' In *Governing the energy challenge: Canada and Germany in a multi-level regional and global context*, ed. Burkard Eberlein and G. Bruce Doern. Toronto: University of Toronto Press, 71–98.

Doern, G. Bruce, ed. 2005. *Canadian energy policy and the struggle for sustainable development*. Toronto: University of Toronto Press.

Homsey, Joe. 2008. 'The evolution of Ontario's emissions reduction trading system.' Master's thesis, York University, Faculty of Environmental Studies.

Independent Electricity Market Operator. 2002. *Monthly market report: July 2002*. Toronto. http://www.ieso.ca/imoweb/pubs/marketReports/monthly/2002jul.pdf (accessed 2 August 2005).

Ontario. Market Design Committee. 1999. *Final report of the Market Design Committee.* January.

Ontario. Ministry of Energy and Infrastructure. 2009. Ontario's coal phase out plan, news release, 3 September. http://www.news.ontario.ca/mei/en/2009/09/ontarios-coal-phase-out-plan.html.

Ontario Power Authority. 2005. 'Overview of the development of power system planning in Ontario.' In Background report 3.1 to the Supply mix report at http://www.powerauthority.on.ca/Page.asp?PageID=122&ContentID=1142&SiteNodeID=127&BL_ExpandID= (accessed 13 February 2009).

Ontario Power Generation. 2001. 'Ontario Power Generation greenhouse gas action plan – 2000.' October.

Trebilcock, Michael J., and Roy Hrab. 2005. 'Electricity restructuring in Ontario.' *The Energy Journal* 26, no. 1: 123–46.

Wikipedia. 'Ontario electricity policy.' http://en.wikipedia.org/wiki/Ontario_electricity_policy (accessed 6 May 2009).

PART ONE

A Global Perspective on Electricity Policy for Ontario

2 Ontario Electricity Policy: The Climate Change Challenge

EDWARD A. PARSON

This chapter looks ahead, over a time horizon of perhaps fifty years instead of the twenty years of the recently published *Integrated Power System Plan* (Ontario Power Authority 2007). Issues developing over this longer time period are relevant to near-term decision making because the long lifetimes of capital and infrastructure in the electrical sector mean that long-term uncertainties push back to affect near-term planning and decisions. Over this fifty-year time period, electricity policy in Ontario will face many challenges, but the most serious – or to be more precise, the most serious that we can now foresee – is global climate change.

Although there are many scientific uncertainties about climate change, as there are in every area of continuing scientific inquiry, the basic scientific knowledge that establishes the seriousness of the risk is largely beyond dispute. As summarized in the most recent assessment report of the Intergovernmental Panel on Climate Change (2007), the Earth's climate is warming, and especially rapidly over the past three decades or so. Human emissions of greenhouse gases – principally the carbon dioxide (CO_2) released by burning fossil fuels, plus several other gases – are predominantly responsible for the observed warming. The climate system is also subject to various natural influences and internal variability, which account for the climate variations that occurred before the gradual emergence of human influence over the past century or so, but human forcings are now the dominant driver of climate change. Under these forcings, climate change will continue and probably accelerate over coming decades. Further global warming is projected to range from 1.1°C to 6.4°C by the end of this century, that is, from roughly 50 per cent greater than the warming of the twentieth century to ten times greater. This wide range of projections reflects both uncertainty about future trends in emissions and other human-driven climate forces, and uncertainties about how sensitively the climate system responds to these forces. Consequently, to track the bottom of this range – that is, to see 'only' 50% more warming in this century than in the twentieth century – would require that *both* emission growth and climate sensitivity lie right at the bottom of the range of current projections. If we do not win both these long-shot bets, we face faster warming, possibly much faster.

Less is known about the impacts of climate change than about the basic atmospheric processes driving change. In some regions, such as the Arctic, the impacts of the climate change that has already occurred are obvious, even acute. As climate change continues, its impacts will be felt everywhere, in widely varying ways, some expected and some unexpected, and their severity will range from irritating to alarming to devastating. At present, the best guess appears to be that climate-change impacts will be disruptive and costly but far from catastrophic – *if*, and this is a big *if*, three conditions hold. This projection only applies for regions that are rich, well governed, and lie in temperate latitudes; it only applies through late this century, not beyond; and it only applies if the realized rate of climate change lies somewhere in the middle to bottom of the range of current projections, not at or near the top. Drop any of these three conditions – that is, for those who live under conditions of less good fortune or greater climate sensitivity, or if climate change lies near the high end of current projections, or if we consider impacts beyond the end of this century – and the possibility of much more severe impacts rises.

The upshot is that dealing with climate change will be on top of political agendas for decades. Unless society utterly fails to come to grips with the climate challenge, we will live in a tightly carbon-constrained world within a few decades, by the end of the Integrated Power System Plan (IPSP) or soon thereafter. Canada will very likely be compelled to take action to reduce greenhouse gas emissions. The biggest questions about this action are not whether or not it will happen, but only whether it will be timely, rationally conceived, and cost-effective or belated, reactive, and excessively costly; and whether it will be chosen by Canadians or imposed by diplomatic and market pressures from elsewhere.

A Carbon-Constrained World: How Big a Change and at What Cost?

How much must emissions be reduced to manage climate change risks, and how much will it cost? Several recent assessments, including the new IPCC report and an analysis done for the U.S. Climate Change Science Program (CCSP) (Clarke et al. 2007), are converging on similar answers. The size of the required changes can be demonstrated by working backwards step by step from the environmental goal. Each step of this reasoning embeds some uncertainties and assumptions, of course, but these numbers give a fair sense of where current judgments fall. Many analysts have suggested that avoiding the worst risks of climate change will require limiting global average warming to about 2°C. Current uncertainty analyses suggest that to have a fifty-fifty chance of meeting that 2°C limit requires holding atmospheric concentration of greenhouse gases to the equivalent of about 450 parts per million (ppm) of CO_2.[1] Holding concentrations to that limit probably requires cutting world emissions of greenhouse gases by between 50% and 85% from present levels by the middle of the century. Finally, plausible assumptions about how soon the major developing countries might agree to deflect their emissions growth suggest that the cuts by rich industrialized countries must be at least 80% by mid-century.

Achieving emission reductions of this scale will require a complete transformation of the world's energy system over the next several decades. Although the required changes are extreme, the fact that they can be implemented over several decades holds the possibility of limiting costs by developing new, low-emitting technologies and phasing in new capital equipment as the old is retired. Moreover, it appears likely that the required emissions reductions can be achieved through technological changes in how society gets and uses energy, rather than by requiring fundamental changes in lifestyle or social organization. This improves the prospects for managing climate change.[2]

Still, the estimates of how costly and difficult the required changes will be span a wide range. The recent analysis of stabilization scenarios for the U.S. CCSP illustrates the present range of judgments. This analysis used three respected energy economic models to estimate costs of stabilizing atmospheric greenhouse gases at four levels, the lowest stabilizing CO_2 at 450 ppm.[3] Two models calculated the total cost of stabilizing at this level as 1% to 2% loss of future gross domestic product (GDP) at a marginal cost of $200 to $400 per ton carbon equivalent, while the third model estimated about 15% future GDP loss at a marginal cost of $2,000 to $3,000 per ton. The key factors producing this wide discrepancy between the first two models and the third model were assumptions about how easily the economy could substitute carbon-based energy with other inputs. In addition to projecting higher economic growth without emissions constraint, the high-cost model assumed more difficult substitution between carbon-based energy and other factors of production, less responsiveness of emissions-reducing innovation to price and policy signals, and strict political constraints on nuclear power. This range of estimates represents real uncertainty in the cost of limiting emissions; while the modellers made different judgments about this uncertainty, current knowledge provides no basis for rejecting any of these judgments as implausible.

Climate Change and Climate-Change Policy: The Outline of a Sensible Response

Given the magnitude of the challenges, it is necessary to make this chapter's opening statement more precise. The biggest fifty-year challenge to Ontario electricity is likely to be the need to sharply cut greenhouse gas emissions, driven by climate change *policy*. This is not to discount the challenges posed directly by climate change and its impacts, which are also serious. The electrical sector is sensitive to climate change through shifting demands (especially for heating and cooling loads), the risks from extreme weather events to transmission and other vulnerable infrastructure, and, especially, hydrological effects on supply. This last vulnerability is obvious for hydroelectric generation but is not limited to that resource. A striking example of unanticipated electric system vulnerability to climate change came in the 1988 summer drought and heat wave. In the central U.S. states along the Mississippi River, this drought caused simultaneous losses of electrical supply from hydroelectric, nuclear, and coal plants. The cause for loss of hydro supply is obvious – low reservoir levels – but the losses of nuclear and coal supply were quite unexpected, although obvious in retrospect. Nuclear supply was lost because river flows fell too low to meet emergency cooling requirements in case of accident. Coal supply was lost because coal barges could not navigate in the low waters. Still, despite this and other examples of vulnerability to climate change impacts, I contend that for Ontario electricity policy these are likely to be secondary to the challenges posed by climate change policy.

To flesh out the nature of this challenge, we must consider some specifics about potential climate change policy. Initially, I will operate on the assumption that climate change policy is crafted to be sensible – that is, effective at addressing the problem, efficient in limiting costs, and intelligently designed so as to be politically feasible and sustainable. What would such a policy look like?

Although any sensible climate change strategy must include measures to adapt to climate impacts, the policies that matter more to electricity policy and planning will be mitigation measures: policies to promote the emissions cuts and transformation of the energy system discussed above. An effective mitigation strategy requires three elements. First, policy instruments are needed that motivate emission reductions by putting a price on emissions, that is, making it costly to emit. These market-based policy instruments can take the form of emissions taxes, systems of tradable emissions permits (also known as 'cap-and-trade' systems), or blends of these two approaches. Second, additional, more focused measures are needed for sectors or technologies that have high impacts on emissions but where the economy-wide, market-based measures are unlikely to be sufficiently effective. These may include more narrowly targeted regulations on high-impact sectors such as buildings and transportation – either additional market-based measures or well-crafted, performance-based conventional regulations, such as efficiency-based reforms to building codes or efficiency regulations for vehicles, major appliances, and other high-impact equipment. They may also include systematic attention to reducing greenhouse gas emissions in connection with major policies and decisions that necessarily fall under the authority of governments, such as planning, zoning, and permitting, and public investment in infrastructure or other major development projects. Third, measures are required to promote research, development, and demonstration projects for key climate-safe technologies. These can also take various forms but must include public support for early-stage, high-risk technologies that promise large contributions to emission reductions, as well as economic opportunities in a future, carbon-constrained world.

These three elements are the building blocks of mitigation strategy. There are many ways to flesh out each of these elements to design and implement actual policies and to combine them into a complete mitigation strategy. In the process there are plenty of ways to construct foolish, excessively costly, and ineffective strategies, just as there are to build effective, efficient, and sensible ones. To help increase the likelihood of mitigation strategies of the latter type, the following principles provide guidance for policy design.

- *Markets where they work, but not where they don't:* Market-based policies leave to citizens and firms greater flexibility in deciding how to achieve emission reductions, and generally achieve environmental goals at lower costs. Consequently, they should be favoured wherever they can be effective. However, the choice of a mixture of market and non-market measures in a complete strategy must be guided by a pragmatic concern for effectiveness and cost, not by ideology. Where efficient, transparent markets cannot be constructed, or market-based policies cannot motivate or coordinate the required changes, other measures must be employed.
- *Broad and consistent incentives:* Where market-based measures are used, they should be applied as broadly as possible across the entire economy in order to give consistent incentives to reduce all emissions from all sources – in particular, to motivate changes and innovations in areas that policy designers could not have identified in advance.
- *Appropriate timing*: To achieve the required multi-decade transformation at manageable cost, it is imperative that the stringency of policies – for example, on the level of carbon taxes, emissions caps, or performance standards – be varied over time. They should start gently but firmly; that is, they should be strong enough to get decision makers' attention but not so strong as to provoke massive premature abandonment of existing capital. Thereafter, their stringency should increase over time as needed, avoiding abrupt changes and providing ample notice of changes in order to smooth investment planning decisions.
- *Technology neutrality*: As we do not know in advance what mix of low and non-emitting technologies will be most effective at achieving the required reductions, the major policies to promote reductions should not tilt the field in favour of or against any technology that holds the promise of making a contribution. The exception to this principle is that measures to promote and support climate-safe research, development, and demonstration (RD&D) cannot avoid

making bets on specific approaches and technologies in deciding what particular projects to support. Even here, however, to the extent that multiple promising avenues are available, these measures should also support a portfolio of alternative climate-safe options, not just one.

- *Fair distribution of burdens:* While it is important to maximize the economic opportunities offered by climate change policies, it is also essential to acknowledge that there will be burdens and losers and to compensate those hit the hardest. There are many dimensions to the distributive effect of policies, for example, among economic sectors, regions, and individuals, but in Canadian greenhouse gas mitigation the most crucial dimension is finding some way to make the burdens of national greenhouse gas mitigation acceptable to the provinces with highest per capita emissions, Alberta and Saskatchewan.
- *International coordination*: Canada's response must be coordinated with action by our major trading partners, yet somehow avoid the current deadlock in which everyone is waiting for others to take the first step. The balance of leadership and international coordination is a perennial dilemma for Canada: a review of Canadian international environmental policy that I authored several years ago characterized it as an attempt at 'leading while keeping in step' (Parson et al. 2001). When Canadian actions do move ahead of those of our major trade partners, it is crucial to manage the resultant competitive effects. Among other things, this means not being afraid to use trade-related measures, such as border tax adjustments or the equivalent, to keep these competitive effects manageable.
- *Keep it simple*: In contrast to the preceding principles, which resemble others that have been widely endorsed, this one, while crucial, has been neglected. Many proposed climate change policies are so complex they look like the tax code. This complexity raises serious problems that have thus far been overlooked. Complex and opaque policies risk diverting effort into clever means to profit by gaming the system, rather than motivating the long-term research and investments needed to cut emissions. Complex policies also make it hard to monitor and evaluate performance. The key to motivating the required changes, and to being able to hold decision makers accountable, is to keep policies as simple and transparent as possible.

Implications for the Electrical Sector: Future Generation Sources

Suppose Canada were to enact a serious and sensibly designed set of policies to reduce greenhouse gas emissions, as sketched above. What would the implications be for Ontario electricity policy? The question can usefully be divided into two parts: how much electricity will be required, and how will it be generated? Although these are linked, I will initially consider them separately and will address the second part first.

What would a serious greenhouse gas mitigation strategy mean for the mix of sources used to generate Ontario's electricity over the next several decades? On this point, recent analyses have been strikingly consistent. If mitigation is pursued through a strategy that includes policies to put a consistent price on emissions across the economy, then atmospheric stabilization at levels near 450 ppm of CO_2 equivalent implies a shift to total or near-total decarbonization of electrical supply, worldwide, by mid-century (Clarke et al. 2007).

The broad classes of technology available for carbon-free electrical generation are well known. The major alternatives for Ontario to consider are hydroelectric power; decentralized renewable sources such as wind and solar power; carbon capture and sequestration; and nuclear power. The good news is that Ontario is moderately fortunate in its endowment of climate-safe resources. The bad news is that none of these alternative sources of supply is free of problems and challenges. All these sources can be expanded, but none of them avoids hard choices or makes the problem of limiting emissions easy.

Hydroelectric energy has been a large share of Ontario's generation since the early twentieth century. Although most of the best sites have been built, it is still possible to build more (as planned in the IPSP) provided local opposition to siting can be overcome. Ontario does not, however, have the option of having an all hydroelectric system like Quebec, Manitoba, and British Columbia have. One alternative to building more hydro in Ontario would be to negotiate large, long-term imports from hydro-rich neighbouring provinces. This approach would still require being able to build the required transmission – again surmounting local opposition to siting – and also paying the price, since these provinces' alternative to selling to Ontario is selling high-priced exports into American markets.

Further expansion of decentralized renewable sources such as wind, solar, and perhaps biofuels is also attractive, but it is crucial not to be naive about the implications. A big capacity build-up means a big environmental impact, no matter what the source. Moreover, even for renewable energy it is necessary to take on siting challenges, both for the generation and for the transmission needed to carry energy from these decentralized resources to major load centres. Finally, and unsurprisingly, manufacturers cannot keep up with the present demand for both wind and solar installations. A big build-up of these would mean paying big rents to foreign suppliers in current tight market conditions, because Canadian firms are not major players in these technologies.

Carbon capture and sequestration (CCS) technologies allow use of fossil fuels without releasing CO_2 to the atmosphere, through various approaches that separate a hydrogen-rich stream for combustion (emitting mainly water vapour) from a carbon-rich stream to be placed in long-term stable sequestration sites underground. In view of the stated political commitment to close all coal generation, consideration of CCS technologies may be moot for Ontario, but if other proposed sources fail to deliver the required expansion, these technologies may have to be re-examined. Indeed, the sustainability of the commitment to shut down coal has to be considered less than certain. Recent history and the continued availability of coal as swing capacity both suggest it will be hard to deliver on the commitment unless all the other pieces fall into place. Certain aspects of the recent attitude to coal, moreover, are clearly not sustainable. It is not sustainable to keep existing plants running full blast, without adequate maintenance or installation of modern environmental controls, under cover of the declaration, 'It's okay; we're closing them in a few years.' Nor is it sustainable to replace coal generation in Ontario by imports from other coal-based systems and claim that it is then the other system's problem. Neither of these stances represents progress towards a climate-safe electrical system.

In any case, Ontario is not especially well endowed in good sequestration sites. Those that have been identified are relatively small and lie in the far north and along the U.S. border, in contrast with the situation of Alberta and Saskatchewan, which have large sequestration capacity in the same sedimentary formations that have provided their fossil-fuel resources. International aspects of CCS are likely to be especially difficult for Ontario. From my perspective, teaching international environmental law in Michigan, I see several student papers every semester expressing outrage at Toronto's garbage coming to Michigan landfills. If CCS in presently identified border sites is seriously considered, I can foresee a good slogan for the next Michigan election campaign: 'We've been taking Toronto's garbage for years, and now they want to send us pollution from their dirty power plants too.' This is not to say that CCS is impossible for Ontario, only that it poses severe difficulties and may not be worth the effort, given the small size of the resource.

Finally, we come to nuclear power. For Ontario, nuclear power poses the biggest and most difficult choice in the shift to non-carbon electricity supply. Clearly an available, non-carbon technology, nuclear power still faces challenges of safety, waste disposal, security against attack, and contribution to nuclear weapons proliferation. The difficulties of nuclear power are clearly evident in the ambivalence of the IPSP. It recognizes that Ontario needs nuclear power, and includes refurbishing and rebuilding reactors to bring nuclear capacity back to the 14 gigawatts (Gw) of the 1990s. At the same time, it proposes to cap total nuclear capacity at that level, with no building on new sites. This skittishness is understandable. Nuclear remains the greatest hot-button issue of energy policy, with many opponents who have not squarely faced the size of the challenge to reduce greenhouse gas emissions.

Moreover, the problems of nuclear power are not limited to the difficulty of making a large-scale expansion politically acceptable. Vendors are marketing new designs that are better than those of the older generation (developed in the 1960s and 1970s) in cost, reliability, and safety. However, it is not clear that the advances achieved are sufficient for acceptability of the huge build-up that has already begun, and that is widely expected to continue, in the major developing countries. Several key concerns remain: the amount of progress that current designs have actually achieved toward 'inherent safety,' which would make core-melt accidents with release of radioactive material essentially impossible; protection against sabotage, attack, or the diversion of fissile material to weapons use from power plants or other more vulnerable points in the fuel cycle; the acute vulnerabilities from continuing to store spent fuel in reactor-side pools, while awaiting development of long-term disposal sites; the proper establishment of a culture of safety and accountability in organizations responsible for building and operating nuclear plants, particularly at senior management levels in market-driven, deregulated systems; and finally, the continued weakness of regulatory oversight, and cosy relationships between industry players and watchdogs.

In view of all these concerns, it is not clear whether the industry is where it must be for a large-scale expansion to be economically viable, publicly acceptable, and safe. Perhaps the scenario of greatest concern is that a rapid, sloppy build-up proceeds worldwide, with inadequate oversight, until another significant accident – a Chernobyl, or perhaps only a Three Mile Island – seizes the public imagination and provokes a backlash so severe that the nuclear option is taken off the table. Even if only a few large economies abandoned nuclear power, such a removal of one of the major options could make it politically and economically impossible to cut emissions enough to avoid serious climate change risks.

Beyond the general challenges posed by worldwide nuclear power development, there are additional tough choices in Canada, particularly in Ontario, associated with Canada's indigenous technology CANDU. The CANDU supply chain has been limping along during the past few decades, thinly nourished by rebuilds of Ontario's reactors and a few export sales. CANDU is a unique technology, differing from other designs in substantial ways that carry both advantages and disadvantages – for example, the use of unenriched uranium fuel (changed to slightly enriched fuel in the recent design tweak), the need for heavy water (formerly as both moderator and coolant; in the recent design change, only as moderator), the advantage of online refuelling, and (until the recent design change) the disadvantage of a positive void coefficient, which probably made earlier CANDU designs unlicensable in the United States. After thirty years in the wilderness – albeit in the wilderness together with the other major nuclear technologies – it is an open question whether or not CANDU has the human, technical, and financial resources behind it to win sales, build safe, reliable, cost-effective new plants, and keep up in the coming design race. I cannot presume to know the answer to this. It remains possible that CANDU could compete successfully in world markets, but this is by no means assured; the technology remains what it has always been, a large technological and business gamble for its promoters and buyers.

The problem of what to do with CANDU is bigger than Ontario, but Ontario has a special place as the principal home of the Canadian nuclear industry and as the province that most needs nuclear power. The fate of CANDU will probably be decided here; it is hard to imagine that CANDU can succeed in world markets if it cannot win large contracts in Ontario. A decision to pursue CANDU would likely mean a national commitment to full support of the technology for reasons of national industrial strategy. Given the stakes in the coming nuclear expansion and the requirements for competing effectively in world markets, the only policy choice that appear to make sense is either such full support or abandonment; continued half-hearted support looks like the worst choice, draining talent and resources without providing a decent shot at success. On this point, it is important to note that abandoning CANDU would not mean abandoning Canadian participation in nuclear power; while CANDU is Canada's only indigenous nuclear reactor technology, Canadian firms are already major participants in other technologies, including some of the most exciting new ones.

The decision of whether or not to make another large commitment to CANDU poses a stark instance of what Albert Hirschman (1970) called the 'Exit and Voice' problem. Hirschman proposed an explanation of why so often one cannot get others (for example, businesses, schools, or government agencies) to provide good performance or good service, even when it appears that one has both the authority to demand it and the option of going elsewhere. In the CANDU case, potential buyers of the technology – in Ontario and elsewhere – need to motivate the highest level of sustained technical and cost performance from the creaky CANDU apparatus. The most effective way to do this is to maintain the threat of obtaining contracts elsewhere. However, since a substantial CANDU building program would in all likelihood mean a national commitment to the technology for reasons of industrial strategy – not necessarily a bad thing – it would be hard to keep this threat credible enough to motivate the desired level of performance.

This is a very hard problem. An indication of how hard it is can be found in a recent expert panel report on energy technology convened by Natural Resources Canada (NRCAN) (2006). This report exhibited not just good judgment but also admirable cogency and clarity in its recommendations on all technologies except nuclear. When they came to nuclear, their text suddenly became uncharacteristically opaque and committee-like, and then they punted, calling for another independent expert panel dedicated to just this question. Perhaps this is the best that can be done at present. The fate of CANDU is a high-order political question, to be addressed jointly by the federal government and the provinces that are the largest potential domestic buyers, principally Ontario. Although the large-scale political dimensions of this choice, and the gamble it represents, cannot be avoided, perhaps a necessary first step is one more attempt at dealing with the technical aspects of the question by a truly impartial, first-rank expert panel.

Implications for the Electrical Sector: How Much Electricity?

The question of how much electricity will be needed defines the envelope within which decisions about the particular mix of generating resources must be made. To consider how emissions limits, and the climate change policies to achieve them, may affect total electricity demand, I go back to the recent CCSP scenarios exercise. In this analysis, the strictest stabilization scenario reduced world primary energy consumption sharply from baseline levels, by roughly 20% to 50% by 2100 in the different models. Electricity production fell by much less, however, from only a few percent to about 25% below the 2100 baselines. This was because the electric share of total energy increased substantially, as price-induced increases in the efficiency of electricity use were outweighed by general economic expansion and the electrification of new energy uses. The result was that even in the strictest (450 ppm) stabilization scenario, electricity generation grew from the present 50 exajoules (EJ) per year worldwide to 150–300 EJ in different model runs. More important, stabilization sharply increased the uncertainty in future electrical demand: electric supply in models' baseline scenarios ranged from 230 to 320 EJ per year, while in the strictest stabilization scenario it ranged from 150 to 300 EJ.

Stabilization also raised the cost of electricity, by 20 to 100 percent above the model baselines.[4] Price increases for electricity were smaller than for transport fuel because it is easier to switch to non-carbon sources in the electrical sector, but even here stabilization will mean higher prices and more price uncertainty. Higher prices were needed to motivate both strong increases in conservation and efficiency improvements and the deployment of new, non-emitting generation technologies. The biggest price spikes occurred in all scenarios in mid-century, during a period of tight capacity as new technologies were coming online but not keeping up with growing demand.

Electricity policy and planning is difficult even without the added challenge of climate change policy. The long lead times of electricity planning create vulnerabilities to uncertainties in demand, policy, and technologies. Running a reliable and low-cost system requires adequate demand forecasting, including forecasting over the long horizons needed to bring major new generation and transmission online. Such long-term forecasts are notoriously prone to the error of overgeneralizing from recent history – a phenomenon with which Ontario has had plenty of unfortunate experience, in its recent history of a decade of building too much followed by a decade of building too little.

Even sensible and well-crafted climate policy, such as a pre-announced thirty-year trajectory of gradually increasing carbon prices, will substantially increase the uncertainty of these demand forecasts by introducing two new effects, each potentially large, with opposite signs. Conservation and efficiency improvements, motivated by prices and policies, will decrease electrical demand, while electrification of new end-uses will increase demand. The electrification of new end-uses includes the elephant in the room, transportation. How many battery-electric vehicles and plug-in hybrids must the Ontario electrical system be prepared to power in twenty years and in fifty years? These two effects, both uncertain, both potentially large, and offsetting to an uncertain degree, will greatly increase the overall uncertainty in demand forecasting. The more uncertain overall demand forecasts become, the more pressure there will be on any particular supply plan, and on price – as the principal tool to match supply and demand.

What about Bad Policy? Potential Climate-Policy Failure Modes and Their Implications

Up to this point I have considered the hard challenges to Ontario electricity posed by a climate change policy that is serious, sensible, and well designed and implemented, but surely we cannot assume that it will be. The response to climate change is already a high-stakes political issue and will remain so for decades. As such, it is likely to exhibit the same mess of real and phony alarms, real and phony solutions, wisdom and folly, and civic leadership and crass advantage-seeking that are perennially present in high-stakes political issues. Even worse, climate change will represent a huge change in energy policy, and large policy changes are especially prone to big mistakes, big attempts at rent-seeking, and big mid-course corrections. There is little reason to expect that the huge transformation of energy policy represented by climate change will be undertaken with more consistency and good sense than are other major policy changes – such as, to pick one recent and relevant example, policies to restructure electricity markets and introduce competition.

What if the Ontario electrical sector has to live with bad climate change policy? I will propose a few salient ways that climate change policy might go wrong, and sketch the implications of each for the electrical sector. (This can make an amusing parlour game: identify plausible ways that policy might go bad, and reason through their implications. I urge the reader to join in the fun.)

Four failure modes strike me as most likely. The first might be called 'the phony war' or 'wait, then hurry up.' This represents a simple extrapolation, perhaps even the most likely one, of the path we are currently on. In this scenario, policymakers do nothing effective to reduce greenhouse gases – perhaps literally nothing or, more likely, a continuation of the present disorganized collection of ineffective and largely symbolic measures – for another twenty or thirty years. Then at some future date, based on some change in public concern or political opportunity, or perhaps after some extreme climate-related event finally galvanizes public attention, policymakers embark on a war-scale effort to cut emissions, spending 5% or 10% of GDP on the effort.

For the electrical sector, the effect of a long 'phony war' would be that climate-safe technologies would not get the research, development, demonstration, time, investment, assessment, or the development of public acceptability that they need to be ready for large-scale deployment. Consequently, when later called on for rapid expansion, these technologies would be more likely to be immature, unreliable, excessively costly, and perhaps unsafe. Moreover, the possibility of abrupt future changes in policy stringency would further increase uncertainties in forecasting overall demand. The cautionary tale of the North American automakers today is relevant, especially that of Ford Motor Company. Bill Ford tried to turn Ford around to a commitment to sustainability nearly ten years ago. He was absolutely right about what changes had to be made, and he utterly failed to move the organization. If he had succeeded, Ford would be in the strongest competitive position of any major automaker today, instead of neck and neck with General Motors for being most at risk of bankruptcy or foreign take-over. Orderly development and roll-out of new climate-safe technologies takes time, whether in transport or electricity, and this can best be provided by gradually increasing the stringency of policies, starting today.

The second failure mode would be an insistence that energy prices must not rise. This could occur overtly if regulators or political authorities simply refuse to tolerate price increases, particularly at the retail level, or in the slightly subtler form of refusing to invest in anything other than least-cost sources – even with the large uncertainty now hanging over future carbon prices. The poster children for this failure mode are a series of recent decisions rejecting or cancelling proposed carbon-capture and sequestration plants because – unsurprisingly – they would have cost more than conventional coal plants. In both Saskatchewan and Wisconsin, proposals to build CCS plants were abandoned after regulators refused to let plant operators pass along the extra cost in electricity prices. Several proposed CCS demonstrations in the United States have also been cancelled, for the stated reason (among others) of cost escalation. In a sharp irony, these cost escalations have been substantially influenced by current high energy costs.

The effects of this policy failure would be similar to those typical of price controls – still more demand uncertainty, plus the risk of shortages and system unreliability as generating capacity falls below growing demand. The resultant disequilibrium would appear either as price gaps between the supply and demand sides that must somehow be covered (presumably through more taxpayer-financed bailouts) or as physical shortages that must be resolved through non-price mechanisms. The available options for doing so – for example, rationing, or prohibition on new electrical hook-ups – are not attractive.

The last two policy failure modes I will consider are a pair; they might be called 'too much market' and 'not enough market.' Having too much market would mean pretending that creating emissions markets will make the required cuts easy, avoiding costs or hard risky choices. This failure is foreshadowed in today's excessive enthusiasm for offsets, which appear to be a cheap and painless way to reduce emissions – 'Just $11.20 to offset your emissions from a round-trip flight between Toronto and Vancouver.' Excessive reliance on markets, particularly complex markets in environmental securities, risks distracting people with chasing short-term gains in emissions trading (instead of pursuing the investments and innovations needed to actually reduce emissions) and driving up price volatility (so the sustained clear incentives needed to motivate these investments and innovations are not present). The first phase of the European Union's Greenhouse Gas Emission Trading System has provided a cautionary example of both these risks.

However, it is also possible to not have enough market. A climate strategy with too little market would not let decentralized market choices determine the particular mix of climate-safe technologies to be developed and deployed, and instead would try to specify the preferred technology mix through political or administrative decisions. Even if the choice of technologies reflects competent and unbiased technical judgments, the failure to subject technology and investment choices to market disciplines carries increased risks, which are borne by the public (whether as ratepayers or taxpayers) rather than private investors. Moreover, such decisions risk being driven by fashion or by special pleading for and against particular technologies or, even worse, risk changing favoured technologies every few years. This approach would raise several grave risks: stranded assets; investments deterred by the expectation that they will be stranded by some future policy change; and powerful incentives for rent-seeking to promote favoured technologies. The largest risk, of course, is basing system planning on favoured technologies that fail to deliver.

Conclusions: Climate Change Issues for Ontario Electricity Policy

In conclusion, I identify seven major issues and challenges to be considered in shaping Ontario's electricity policy for the coming climate-constrained world.

1. Ontario electricity policy planning must consider the impact of climate change – direct climate impacts to be sure but, even more important, the impact of potential climate change policies. As those responsible for electricity policy decisions will usually not be those who are controlling climate change policy, it is essential to consider the implications both of well-crafted policies and of the highest-risk forms of bad policy. Planning may greatly benefit from scenario exercises that focus on these implications, although because it is difficult to do these in a way that both usefully sketches out the implications of major policy risks and is sufficiently tactful, they may need to be done confidentially.
2. Ontario is relatively fortunate in the mix of generating resources available to it in a climate-constrained world. As the IPSP shows, one can construct a system plan that has 90% carbon-free energy by 2027 (with only 10% natural gas remaining), cutting greenhouse gas emissions by more than 80%, if all goes well. However, even being this carbon free – for that matter, even planning on being 100% carbon free – does not avoid all risk from climate change or climate change policy. The reason is not just the last small slice of fossil generation in the baseline projection but, more crucially, what happens if you need more generation than the baseline anticipates, either owing to faster demand growth or to other planned supply sources not becoming available as expected.
3. Any forceful and serious climate change policy will increase the uncertainty of electrical demand forecasting, even if the policy is ideally designed and implemented. Bad climate change policy means bigger risks – involving supply adequacy, reliability, costs, and greenhouse gas emissions. If one substantially underestimates demand growth, or overestimates availability of nuclear or renewable sources, the system will be less carbon free than expected. It

is imperative to consider a wider range of demand uncertainty for system planning under expected climate change policy.

4. Climate-safe supply technologies are likely to experience a sharp, sustained supply-chain crunch, especially under the 'phony war' scenario at that future time when the phony war becomes a real one. While I am not counselling a return to the extreme demand overestimation and the overbuilding of the 1960s and 1970s, it might be worthwhile to overbuild a little early to manage the risks of higher demand and the associated risks of cost, availability, and lead time of supply.
5. Among other policy levers, price must be allowed to vary. In the long term, limiting emissions requires the prices of electricity, as of all energy, to rise. In the short term, to insulate consumers from short-term price spikes, the system needs extra capacity to call on and the transmission to deliver it. The climate change challenge touches the challenges of electrical market design principally in the need to construct a system that effectively limits rents and self-dealing and provides cost discipline.
6. Maintaining a reliable system to serve a growing economy requires the ability to overcome local opposition to facility siting, no matter what the supply mix. This emphatically includes transmission.
7. Nuclear power poses sharp challenges and hard decisions that decision makers cannot keep avoiding. Ontario may need a substantial expansion of nuclear capacity to reliably hold down greenhouse gas emissions, but this will require progress on several fronts: providing adequate confidence to investors in order to motivate building; solving the problem of regulating effectively in order to provide sufficient pressure for continued improvement in designs and operations; lancing the boil of long-term spent-fuel disposal; and finally, being prepared to make a decision – likely a joint one between federal and provincial governments – on whether or not Canada will continue to support CANDU at sufficient scale to have a chance of succeeding in world markets.

The major actors in Ontario electricity policy, including both the Ontario and the federal governments, must face up to the challenges of running a viable system. These cannot be solved by ideology, slogans, or wishful thinking about problem-free energy sources. This is true even without climate change and policies to limit emissions, but they tighten the screws and make the challenges even harder.

NOTES

1 At the start of the industrial revolution, before we began putting large amounts of additional CO_2 into the atmosphere, its concentration was about 280 ppm. The level today is about 385 ppm, rising by about 2 ppm per year. Other pollutants in the atmosphere, however, complicate the relationship between today's level and any proposed limit. Today, other greenhouse gases emitted by human activities (for example, methane, nitrous oxide, and halocarbons) raise the total warming effect to be equivalent to about 450 ppm of CO_2. Still other pollutants in the atmosphere (sulfates and other aerosols) are exerting a net cooling effect, which brings the total effect of all human-source pollutants back to about the same as 385 ppm of CO_2. In other words, CO_2 alone is at 385 ppm today, other greenhouse gases raise the warming effect to that of 450 ppm of CO_2, while aerosols bring it back down to that of about 385 ppm of CO_2, although with substantial additional uncertainty. These various pollutants all have different residence times in the atmosphere, from a few days or weeks (aerosols) to many thousands of years (some greenhouse gases). If anthropogenic aerosols decrease over coming decades, as seems likely in view of the other environmental harms they cause, then any specified limit on anthropogenic global warming grows more challenging because it must be achieved with less human-driven cooling working in its favour.

2 This is a conjecture, of course. If it turns out to be mistaken or if the required technological innovations cannot be motivated, developed, and deployed at the required scale and speed, it may be that the only way to achieve the required cuts will be through such larger-scale social changes.

3 The four stabilization levels were actually expressed in terms of total anthropogenic radiative forcing, not equivalent CO_2 concentration. The lowest level was 3.4 W/m^2, of which 2.6 W/m^2 came from 450 ppm of CO_2 and the remaining 0.8 W/m^2 from other greenhouse gases.

4 The baseline electricity prices ranged from present levels to about 50% higher, in the different models.

REFERENCES

Clarke L., J. Edmonds, J. Jacoby, H. Pitcher, J. Reilly, and R. Richels. 2007. *Scenarios of greenhouse gas emissions and atmospheric concentrations*. Synthesis and Assessment Product 2.1a. U.S. Climate Change Science Program: Washington, DC.

Hirschman, Albert O. 1970. *Exit, voice, and loyalty*. Harvard University Press: Cambridge, MA.

Intergovernmental Panel on Climate Change (IPCC). 2007. 'Climate change 2007: The physical science basis.'

Natural Resources Canada. 2006. 'Powerful connections: Priorities and directions in energy science and technology in Canada.' Report of the National Advisory Panel on Sustainable Energy Science and Technology.

Ontario Power Authority. 2007. *Integrated Power System Plan*. August.

Parson, E.A., with A.R. Dobell, A. Fenech, D. Munton, and H. Smith. 2001. 'Leading while keeping in step: Canadian management of global atmospheric risks.' In *Learning to manage global environmental risks: A comparative history of social responses to climate change, ozone depletion, and acid rain*, ed. W.C. Clark et al., Social Learning Group, 235–57. Cambridge, MA: MIT Press.

PART TWO

Electricity's Role in Reducing the Environmental Footprint of Energy Use

3 Introduction

HEATHER L. MACLEAN

Electricity policy in the province of Ontario is supported by three pillars: reliability, affordability, and environmental sustainability (Norman 2008). This part of the book is focused on the latter pillar, environmental sustainability.

The environmental sustainability of Ontario's electricity sector is dependent upon the amount of electricity generated within the province, the mix of fuels and technologies utilized to generate this electricity, and the environmental implications of the province's demand-management programs. In addition, the sector's performance is determined by the technologies utilized to generate electricity imported into the province.

Ontario's electricity generation installed capacity is expected to evolve by the year 2025 from the current mix of nuclear (14,000 megawatts [MW]), renewables (8,300 MW), coal (6,400 MW), gas and cogeneration (5,100 MW), as well as a conservation component (1,300 MW) to one that meets the Government of Ontario's Supply Mix Directive of June 2006 (Ontario Ministry of Energy and Infrastructure 2008). According to the directive, the 2025 installed capacity is planned to consist of nuclear (14,000 MW or less), renewables (15,700 MW), gas and cogeneration (9,400 MW), and conservation (6,300 MW). Each of these fuels and technologies has unique impacts on the environment. While some of the technologies may have lesser negative environmental impacts, none of the technologies is benign.

The production of electricity has many benefits and costs. However, the environmental sustainability session of the workshop 'Current Affairs: Perspectives on Electricity Policy for Ontario' focused on greenhouse gas emissions because of the importance of global climate change and the current priority of the associated greenhouse gas emissions reductions in electricity policy. Ontario's electricity generation sector along with the province's industrial and transportation sectors are responsible for the largest shares of Ontario's greenhouse gas emissions. Two components of the province's overall emissions are directly related to the electricity sector: fossil fuel industries, and electricity and heat generation. These components together were responsible for approximately 20% of greenhouse gas emissions in 2004 (Jennings 2008). The emissions of the electricity generation and transportation sectors increased substantially during the period from 1990 to the present, while those in the industrial sector have declined slightly (Go Green Ontario 2007).

The Government of Ontario has a plan for reducing greenhouse gas emissions in the province by the year 2020. This plan states that the largest component of emissions reductions (29%) will result from actions in the electricity sector (Jennings 2008). These include the phase-out of coal-fired electricity in the province, an increase in the utilization of renewable sources, and other electricity policies. It is reported that power plant emissions will be reduced by 85%, from 46 Mt CO_2 equivalent (in 2003) to less than 7 Mt CO_2 equivalent by 2014, when the coal phase-out is planned to be completed (Go Green Ontario 2007). While these reductions will be significant if achieved (and only time will tell if they can be), there are considerable challenges to the Ontario plan. Issues to consider include the increasing demand for electricity, the reliance on natural gas for electricity generation (which also results in greenhouse gas emissions), the success of renewable generation options and conservation programs, imports of electricity to the province (potentially generated from coal) from other jurisdictions, and the full life-cycle implications of electricity generation (the latter point is discussed below).

Some of the key environmental 'footprint' issues that remain to be addressed in the electricity sector include (1) the widely varying environmental performance of electricity generation pathways, (2) attention to sustainability issues, including a broad set of environmental issues as well as social and economic sustainability components, (3) the increasing demand for electricity in Ontario and elsewhere, and (4) the provision of correct price signals by policies addressing environmental issues. While it is beyond the scope of this introduction to discuss each of these in detail, a few key points are mentioned below.

Electricity generation pathways (consisting of fuels and generation technologies) vary widely in their environmental performance. A life-cycle perspective is critical for determining environmental performance. Life-cycle assessment (LCA) examines the implications on the environment of the full set of activities (and their supply chains) associated with the production, use, and end of life of a product (International Organization for Standardization 2006). In the past year there have been several announcements related to incorporating life-cycle-based standards into climate change regulations for transportation fuels. These initiatives will have a direct impact on the electricity generation sector with the increasing electrification of vehicles, making electricity a transportation 'fuel' covered under these regulations (California Office of the Governor, 2007; UK Department of Transport 2006). As well, it is expected that life-cycle-based standards may become more prominent for environmental and climate change regulation and be utilized directly in the electricity sector (and other sectors in the economy) in the future. The most prominent of the life-cycle-based transportation fuel initiatives are California's Low Carbon Fuel Standard (LCFS) and the United Kingdom's Renewable Transportation Fuel Obligation Programme (RTFO) (California Office of the Governor, 2007; UK Department of Transport 2006). Other governments, including Ontario which signed a memorandum of understanding with California, have declared interest in adopting similar standards (Ontario Ministry of Energy and Infrastructure 2008). On 18 January 2007, the State of California, through Executive Order S-1-07, announced the intent to regulate a reduction of at least 10% by 2020 in the life-cycle carbon intensity of transportation fuels sold in the state (California Office of the Governor, 2007). Fuel providers will be required to ensure that the mix of fuels they sell into the California market meets, on average, a declining greenhouse gas emissions standard. The LCFS was adopted on 23 April 2009 and requires that the life-cycle greenhouse gas emissions associated with the production of a large set of transportation fuels (including electricity) be quantified.

All activities associated with the full life cycle of electricity production (the entire supply chain) and use need to be included in determining environmental performance. These activities comprise the extraction of raw materials, transportation activities, electricity generation, transmission, and waste disposal. A life-cycle approach enables comparison of the relative performance of various generation technologies and provides a broader awareness of the overall implications of technologies. In addition, it provides information that can lessen the shifting of environmental burdens from one stage of the life cycle to another or from one environmental medium to another. For example, while electricity generated from nuclear and hydroelectric facilities generally is reported to have few concerns related to greenhouse gas and air pollutant emissions, a broader life-cycle perspective highlights the environmental implications of constructing and maintaining the large infrastructures required for both types of facilities, land use issues, aesthetics, and waste disposal issues (the latter associated primarily with nuclear facilities). While biomass-derived electricity in most cases results in low greenhouse gas emissions owing to the biogenic nature of the carbon in the biomass, considerable fossil fuel and other inputs are utilized in biomass production and transportation (for example, fertilizer inputs, fossil fuels for farm equipment and trucks, and water), and as well there are issues related to land use for many categories of biomass (Zhang, Habibi, and MacLean 2007).

Sustainability performance is a multi-attribute decision problem. A range of environmental impacts (not solely greenhouse gas emissions) need to be considered as discussed above with regard to life-cycle approaches, along with other factors such as reliability and affordability. Without the broader focus, suboptimal decisions will result.

The increasing demand for electricity in Ontario and elsewhere (including increasing electrification of vehicles) makes it even more critical that we make progress on reducing the environmental footprint and, more broadly, progress towards sustainable electricity generation. Conservation will be key. Finally, enacting policies that provide correct price signals is essential. A range of policy instruments must be examined, and new instruments developed. Lessons from other jurisdictions will be crucial.

With the above context, including an understanding of the scope of the problem and an acknowledgment of the scale of the challenges facing Ontario and other jurisdictions worldwide related to moving to more sustainable electricity provision, three chapters follow that provide insights on different facets of the overall issue of improving the sustainability of the electricity generation sector. The authors discuss the potential of various policies and strategies for mitigating greenhouse gas emissions from the sector. The first chapter examines lessons learned from European experience of low carbon strategies, providing guidance for those in other jurisdictions. The second chapter models various environmental and technology policies and compares their potential with respect to climate change mitigation. The final chapter in this section discusses the approaches of the two prior chapters and raises deeper questions about the role of electricity in reducing carbon dioxide emissions and the nature of effective policies.

REFERENCES

California Office of the Governor. 2007. 'Executive Order S-1-07: The low carbon fuel standard. 18 January.'

Go Green Ontario. 2007. 'Ontario greenhouse gas emissions targets: A technical brief.' 18 June. Available at http://www.ene.gov.on.ca/publications/6793e.pdf.

Government of Ontario. 2008. 'Low carbon fuel standard.' Discussion paper prepared by the Ontario Ministry of Energy. May.

International Organization for Standardization. 2006. *ISO 14041: Environmental management; Life cycle assessment; Goal and scope definition and inventory analysis.* Geneva: ISO.

Jennings, R. 2008. 'Go green and the low carbon fuel standard.' National Conference on Low Carbon Fuel Standards for Canada, 4–5 June. Toronto: Ontario Ministry of Energy and Infrastructure. Available at http://www.pollutionprobe.org/Happening/pdfs/lowcarbonfuelwkshp/jennings.pdf.

Norman, J. 2008. 'Presentation to University of Toronto Leaders of Tomorrow Program,' 23 May. Toronto: Ontario Ministry of Energy and Infrastructure.

Ontario Ministry of Energy and Infrastructure. 2008. 'Backgrounder: Ontario's electricity supply mix.' Toronto: Ontario Ministry of Energy and Infrastructure.

United Kingdom. Department of Transport. 2006. 'Renewable Transport Fuel Obligation Programme (RTFO).' Available at http://www.dft.gov.uk/pgr/roads/environment/rtfo/.

Zhang, Y., S. Habibi, and H.L. MacLean. 2007. 'Environmental and economic evaluation of bioenergy in Ontario, Canada.' *Journal of the Air and Waste Management Association*, no. 57, 919–33. DOI:10.3155/1047-3289.57.8.919.

4 European Low-Carbon Strategies in Liberalized Electricity Industries: Some Lessons on the Efficiency of the Market Paradigm

DOMINIQUE FINON

Socially efficient policies for reducing the environmental footprint of a power system, from generation through to consumption, cannot be developed without taking into account the new market regime in the power industry. There are two reasons:

- In a market regime for a vertically de-integrated system, it is expected that the market price will indicate to producers in a timely way the socially efficient technological choice for generation. However, with a market regime, there tend to be significant investment risks that must be borne exclusively by producers. This discourages the development of clean and non-carbon approaches because they are more capital intensive. Accordingly, to encourage and protect investment it is necessary to implement special arrangements that violate the principles of competition, such as auctioned long-term fixed-price contracts or vertical integration.
- In energy and environmental policymaking, market-based instruments (tax, cap–and-trade instruments, or subsidies) are adopted to internalize the negative (or positive) externalities associated with each activity of the electricity chain. These are consistent with the market regime in the energy sector, as opposed to 'command-and-control' policies involving, for example, emissions norms, energy efficiency measures, closures of polluting facilities, and mandatory development of clean and green technologies. However, market instruments such as cap-and-trade tend to add price and regulatory risks to new market risks, which increases the need for special arrangements for the development of non-emitting technologies.

In contrast, a regulated utility-monopoly regime limits investment risks for producers by enabling cost pass-throughs on tariffs. It also allows programming and strong governmental coordination of generation investment based on integrated resource planning, which takes environmental externalities into consideration.

This chapter focuses on the development of climate policies in electricity industries under market regimes. Their ultimate aim is to reduce CO_2 emissions by encouraging technology shifts in centralized and decentralized generation and by improving efficiency in electricity use, spontaneously reducing the environmental footprint of an electricity system. European low-carbon strategies developed in the electricity industry under the market regime serve as an example. The case of the European Union is instructive, given that the dominant paradigm of regulation of the economy in general, and of complex industries in particular, is the market model; any other types of social coordination for long-term efficiency are perceived as a source of regulatory imperfections. It informs the European Commission and the most market-oriented member states (in particular, the United Kingdom, the Nordic countries, and the Netherlands) in the creation of new legislation for electricity markets and the formulation of environmental policies. European climate policy has been set up as a cap-and-trade regime for the large emitting industries, including electricity industries. Its goal is to design complementary measures to boost the deployment of green technologies (renewables) and promote energy efficiency in a market-oriented way (for instance, tradable green certificates rather than purchase obligations under feed-in tariffs for the renewables in electricity generation).[1]

This chapter has two purposes: to show that an institutional regime in the electricity industry is far from neutral in terms of policy efficiency and to argue that the design of a policy is as important to its social efficiency as are the theoretical principles on which it is based (for example, command and control, norms, tax and subsidy, or quotas in cap and trade). The objectives of the European energy and climate policy to 2030 are presented, followed by the electricity reforms that successive European laws (so-called directives) impose on member states and a discussion of how they contrast with Ontario reforms. This leads to an analysis of the tension between the competition regime proposed for the near future and the environmental policies that require the development of capital-intensive, clean, non-carbon facilities. Next, there is a review of the market-oriented policies implemented or under discussion by the European Union regarding CO_2 emissions reduction and the promotion of renewable production in the market environment. The chapter ends with a summary of the virtues of a hybrid regime, intermediate between the regulated utility regime and the decentralized market regime.

Prospects for the Development of Non-carbon Technologies in the European Union under Market Instruments

In the European Union, more than 50% of electricity generation is based on fossil fuel (more than 30% on coal). According to the new official scenarios of the European Commission (European Commission Directorate-General Energy and Transport 2008), the proportion of fossil fuel generation will decrease steadily with the development of decentralized renewables, large hydro resources are limited, and nuclear development is politically restricted in a number of member states. Emissions from electricity generation could increase by 70% by 2030 if no determined corrective measures are adopted.

Examination of the new baseline scenario of the European Commission through 2030 (which assumes that present policies will be applied in a business-as-usual future) shows that technological changes in power generation may be important because most existing equipment will need to be replaced over the next three decades. This projection does not take into account investment risks. It assumes that capital costs are the same as before the market reform and that there is no risk aversion to investing in capital-intensive equipment; however, as noted, it incorporates political restrictions. The share of nuclear generation (15% in 2005) will drop over the projection period, reaching 10.6% in 2030 (about half of the 2000 share), owing to the incomplete replacement of units to be decommissioned and the phase-out policies followed by certain member states. Combined cycle gas technology (CCGT) will continue its penetration, attaining a share of 23.5% in 2030 because of its moderate level of carbon emissions (half of those from coal generation); steam turbines using fossil fuels (coal, gas, and fuel oil) will have a decreasing share, declining to 30% of total capacity in 2030.

However, market conditions (prices for coal being lower than those for gas and fuel oil) and emergence of supercritical, high-efficiency thermal power generation will enable the limited re-emergence of coal-based generation in the long term. European climate and energy policies could eventually rely on carbon capture and sequestration (CCS), currently in the research, development, and demonstration (RD&D) stage. A recent European Parliament report (Davies 2008) recommends that new coal equipment built after 2015 should be capture ready and that 90% of the CO_2 emissions must be captured and stored by 2025. However, this recommendation is not taken into account in the baseline scenario.

A discussion of renewable energy sources in electricity (RES-E) begins by noting that, owing to the already high exploitation of suitable sites in the European Union, hydropower capacity will expand much less than total capacity. Under the hypothesis of current policies and fuel prices, wind power is expected to grow throughout the planning period, attaining a capacity 3.6 times higher in 2030 than in 2005, which will correspond to 15% of total capacity. The generation capacity of renewable energy, including biomass and waste plants, will account for 34% of total power capacity in 2030, up from 22% in 2005. Decentralized renewable energy capacity will account for 17% of the total in 2030, up considerably from 2% in 2000. The high penetration is driven mainly by the development of wind and biomass power plants. Wind power is projected to attain a capacity of 146 Gw by 2030 (129 Gw onshore and 17 Gw offshore).

However, the 70% increase in carbon emissions in the baseline scenario (that is, without a determined climate policy) is inconsistent with the ambitious objective of reducing the total emissions of the European economy by a factor of four by 2050. In the alternative scenario, which

Figure 4.1. Power generation capacity by type of main fuel used in the EU-25 in Gw

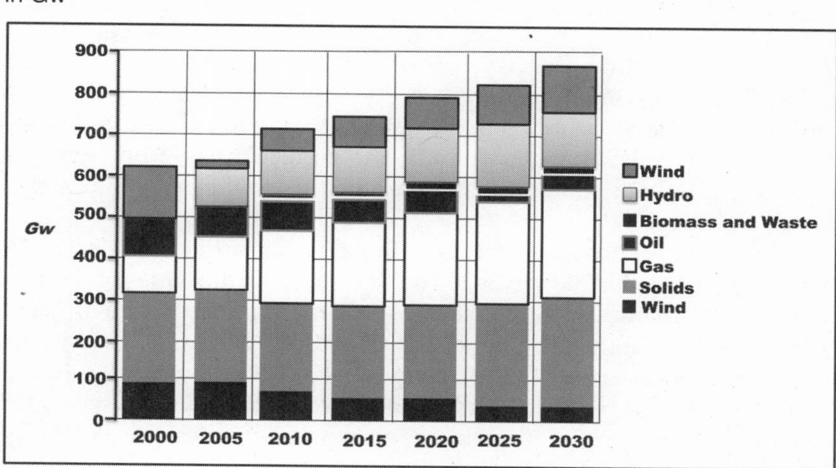

Source: European Energy and Transport – Trends to 2030, Update 2007 (May 2008)

incorporates an increasingly restrictive climate policy, the goal of the European Union will be a 20% overall reduction in emissions between 2005 and 2030. The European Commission evaluated the effects of a market-oriented policy that was designed to reach this target, based on a uniform carbon price, and found that the economic position of emitting technologies in all sectors will be dramatically weakened by the increasing carbon price; it will have to increase in stages to €47 per tonne (t) of CO_2 in 2030 to reach the target.

A comparison of emitting sectors (electricity, industry, transportation, and construction) shows that the electricity industry will have to reduce its emissions by 40% in this stringent scenario because it is more flexible than are the transportation and construction sectors. Nuclear technology and clean coal technologies (CCS) will expand slightly through 2030, constrained by political restrictions and learning processes, respectively, but growth will be more significant during the period 2030–2050.

The European Commission also tested scenarios with complementary policies based on quantitative objectives in renewable energy sources (RES) development and in overall energy efficiency: 20% of the primary energy balance from renewables; 20% reduction in energy intensity. The RES objective is shared among the twenty-seven member states according to resource endowment and the level of prosperity. In electricity generation, this complementary policy increases the share of renewables from 15% in 2005 to 40% in 2030, instead of the 25% in the baseline scenario (350 Gw, instead of 170 Gw, to be installed between 2010 and 2030). It is worth noting that by boosting RES development to a higher level than that reached under the sole influence of carbon price, this policy tends to decrease the demand for CO_2 permits by emitting producers, who would be obliged to develop RES-E technologies at a higher level. Consequently, the policy focusing on RES will depress the CO_2 price that must be reached to limit overall emissions, by €3/t CO_2 in 2030.

To sum up, a European carbon policy that has an ambitious reduction goal (20% from 2005 to 2020) will be structured by market instruments that will impose an increasing carbon price. The price will have to rise progressively, reaching a high of $50/t CO_2 by 2030, but in order to overcome barriers to the deployment of RES technologies, it will be complemented by RES policy focused on quantitative objectives (to reach specified shares of the primary energy balance and the electricity balance).

Contrast with the Ontario Carbon Policy

In Ontario, public policies dedicated to reducing the environmental impact of the electricity system are shaped by the public utility regulation culture. Environmental measures and political decisions, in particular those addressing CO_2, renewables, and energy efficiency, are not based on market-oriented instruments. Examples include

- the decision to eliminate coal generation capacity (6.4 Gw out of 34 Gw, that is, 20% of the installed capacity) before 2014;
- the program for developing new renewables production in large centralized units or small decentralized facilities (an additional 7.4 Gw by 2025);
- the development of new connections with Manitoba and Quebec to import more hydroelectricity as a replacement for emitting technologies; and
- the ambitious electricity conservation program whose target is the equivalent of 5 Gw electricity savings by 2025.

The closure of coal plants is typical of this non-market orientation. In Europe, although directives have been issued for sulphur dioxide (SO_2) and nitrogen oxides (NO_x) emissions regulation, the closure of coal generation facilities has to be encouraged by the allocation of CO_2 permits, either in the form of a free allowance with increasing restrictions or of allocation by auctions, as is the plan after 2012. This approach gives an implicit price to carbon. The market-oriented approach is based on the following logic: imposing a price on carbon affects the decision of every agent at the same time; and consumers of end products pay for the indirect impact of their consumption through the successive internalization of carbon costs along the path of production. Thus, rather than regulatory structures reflecting command-and-control decisions about specific technologies, change will be effected by decisions of decentralized agents which react to price signals after internalization of environmental costs.

Determining the Trajectory of European Electricity Market Reforms

With its successive directives on electricity markets, the European Union attempts to follow the textbook market model for a decentralized electricity market, based on the following principles:

- vertical separation of competitive and regulated monopoly activities to facilitate competition and regulation;
- horizontal restructuring to create an adequate number of competing producers and suppliers in wholesale and retail markets;
- designation of an independent system operator (ISO) to maintain network stability and facilitate competition, supported by market mechanisms for receiving offers of ancillary services (the ISO could be the owner-operator of the transmission system after the system has been completely unbundled from generation and supply activities);
- application of transparent regulatory rules to promote access to the transmission network;
- establishment of market rules for access to distribution networks and to ease competition at the retail level, after suppression of regulated electricity tariffs;
- creation of an independent regulatory agency with adequate staff and powers to implement incentive regulation and promote competition.

In addition to governmental decisions, a power exchange must be developed, either by the regulator and the ISO if it is a mandatory power exchange[2] or by private players. Its role is to facilitate hourly transactions and complement the market with financial contracts. In Europe, wholesale markets are decentralized, and power exchanges complement contractual sales. Whatever the nature of the power exchange, price setting is radically different from that of the historic tariff regulation, which was based on average cost price. Market prices are determined in hourly markets by marginal offers in order to equilibrate supply and demand on the power exchange. They are quite volatile, in particular, during peak periods.

Privatization of former public monopolies may occur after some vertical and horizontal de-integration, and preferably after market liberalization. However, in Europe, as in the rest of the world, the industrial structures of liberalized electricity markets differ significantly, especially in terms of horizontal restructuring to create an adequate number of competing generators, vertical separation of competitive businesses in generation and supply, and unbundling between the distribution network and supply.

A key issue is whether the restructuring of a country's power sector is ideologically and politically motivated, or whether it is forced upon a country by a higher authority, for instance the European Union and its member states, or the World Bank and less developed countries; the latter case often leads to a half-hearted and unstable approach. In Europe, such reforms run afoul of loyalty to public service (for example, in France and Germany) and cultural scepticism about market virtues, in particular, concerning the realization of capital-intensive, large-scale equipment programs (for example, nuclear programs in France). Differences of opinion continue to exist regarding key market design issues – for instance, the choice between a mandatory and a decentralized day-ahead market, the degree of separation of ownership of transmission and distribution network activities from competitive activities, the rules of network access, and the degree to which the industrial structure must be horizontally and vertically de-integrated.

More generally, public acceptance of reform can also be undermined by the price shocks resulting from the transition from regulated tariffs based on average costs to market prices aligned with hourly prices on the volatile power exchange, as occurred in Ontario. In a number of European countries, retail market liberalization, which was supposed to be completed by July 2007, remains severely restricted by price regulation designed to protect consumers, including even medium-sized industrial consumers, who are vulnerable to the first price hook and the effects of wholesale price volatility. Moreover, the development of systems based on an optimal mix of technologies has not been submitted to market tests, owing to overcapacity at the time of the reform. Nonetheless, given aging equipment and new investment cycles in various EU countries, reinvestment in generation is inevitable. In this context, public policies requiring capital-intensive investment and energy conservation actions on electricity demand could conflict with the decentralized market paradigm.[3]

Europe has the opportunity to learn from the U.S. and UK market culture in the design of electricity reforms and in the choice of public policy instruments compatible with the market environment. In contrast to the U.S. and Canadian federal governments, the European Union has the power to impose market liberalization on its member states even if they are reluctant because the raison d'être of the European Union is to stimulate trade and exchange among member states and to pursue economic integration in the context of political integration. Accordingly, directives and *règlements* can impose market rules on national electricity industries, such as the directives of 1996, 2003, and 2008 (European Commission 2003, 2008). The European Commission receives yearly benchmark reports that assess the state of competition in each member state (see table 4.1) and has the mandate to supervise

Table 4.1. Comparison of Horizontal Concentration in European Electric Markets

	Generators with capacity share of >5%	Capacity share of top 3 generators	Retailers with market share of >5%	Market share of top 3 retailers
Austria	5	45%	7	67%[a]
Belgium	2	96%[b]	3	53%
Denmark	3	78%	6	38%
Finland	4	45%	4	33%
France	1	92%	1	90%[c]
Germany	4	64%	3	50%
Greece	1	97%[c]	1	100%[c]
Ireland	1	97%[c]	1	90%[c]
Italy	4	69%	4	72%[b]
Netherlands	6	59%	7	48%
Portugal	3	82%	1	99%[c]
Spain	4	83%	4	94%
Sweden	3	90%	3	47%
UK	8	36%	10	42%

[a]First seven
[b]First two
[c]First one
Source: European Commission, 2005a, *Annual Report on the Implementation of the Gas and Electricity Internal Market*

the application of directives and to propose new legislation to bring competition to the level of the agreed targets.

Although the European Commission (encouraged by the precedent set by the United Kingdom and the Nordic countries) is eager to follow decentralized market principles, the directives are compromises among member states. The reluctance of major players (in particular, Germany, influenced by its two large energy companies, and France, influenced by its technocrats) has produced directives with imperfect and incomplete market rules. The fundamental issue is that the European Commission has no power to impose privatization and vertical or horizontal de-integration unless member states agree on the changes by voting for a directive, such as the new 2008 directive sanctioning the legal unbundling of the transmission system operator. In exceptional circumstances, the European Commission could act on industrial structures in the case of mergers and acquisitions, by requiring divestitures, but competition policy cannot replace industrial policy; the principles of a decentralized market model require the latter.

In fact, the European Commission has not been able to impose a decentralized model. With the support of their governments, historic operators that are now privatized can preserve vertical integration between generation and supply by maintaining ownership of all their generation assets and can also preserve most of their retail market shares, domestic, commercial, and industrial. This has been the pattern in the majority of member states; the most dynamic operators have even acquired regional companies in other member states and in the gas industry.

Nevertheless, environmental and climate policies that require capital-intensive non-emitting equipment and important technological advances, both of which could incur some risk, can be developed in this market environment. Member states that preserve historic utilities and are reluctant to reform their electricity industries could find it advantageous to install large-scale non-carbon equipment (for example, the efficient nuclear plants built by EDF in France) or to implement a costly vertical gas strategy (for example, the international gas infrastructures developed by German players RWE and E.ON in joint ventures with geopolitical players such as the Russian Gazprom).

Contrast with Ontario Electricity Reform

As opposed to member states of the European Union, Canadian provinces have nearly total sovereignty over electricity sector institutions and rules, even those concerning trade between provinces. A province like Ontario typically remains under the influence of the public utility regulation regime, so electricity reforms are subject to the effects of electoral cycles and political instability. In Ontario there has been a succession of restructuring initiatives to dismantle the vertically integrated public monopoly, resulting in a competitive generation sector and liberalizing prices (Trebilcok and Hrab 2006). However, in 2002, confronted with the first price hike three months after the effective kick-off, the government decided to abandon market liberalization. Political consensus led to the rejection of partial or total privatization of Ontario Hydro. Almost all retail prices are defined under a type of cost-of-service regulation in which prices reflect average costs, rather than marginal prices (determined by the marginal offer of the most costly plant), in an hourly market over the year.

In this institutional environment the electricity system is coordinated by the government, which undertakes programming and planning. The creation of the Ontario Power Authority, charged with system planning, rate evolution, conservation policy, and promotion of renewables, institutionalized this style of coordination. The approach reflects a lack of confidence in the ability of market mechanisms to guarantee short-term efficiency and in the prospect that decentralized decisions of players investing according to market price signals will converge with long-term policy.

Tension between Policy Objectives and the Electricity Market Regime

In a decentralized market regime the producers are supposed to bear all the investment risks, while in the old cost-of-service regulation regime the risks were borne mostly by consumers via the pass-through of costs on the tariffs. Financial investors and pure producers have a preference for technologies that facilitate good market risk management through the correlation of electricity prices to fuel costs. However, in a non-carbon project requiring large capital investment it is impossible to manage market risk in this way.

Unfortunately, the market regime may not be compatible with climate policy objectives and, more generally, with other converging energy policy goals (energy security, energy reliability, and industrial and employment policy) affected by these technologies. This raises several crucial issues: How can the development of low-carbon or non-carbon generation equipment be encouraged? Should the production of new, non-carbon facilities be taken out of the market and allowed to benefit from the security of vertical arrangements that enhance the prospects of their development? Do integrated and large-scale companies provide advantages for developing investment strategies that can converge with policy goals?

Capital-Intensive Investment in Risky Electricity Markets

Decentralized markets involve sell-side risks arising from highly fluctuating power prices in the absence of storage and from the price inelasticity of real-time demand. Specific to the power sector, these risks discourage the development of long-term hedging products and increase the reluctance of large wholesale buyers (suppliers/retailers, major consumers) to contract on a long-term basis and share investment risks with new producers.

A fundamental aspect of risk management that handicaps capital-intensive investment decisions in de-integrated electricity industries is the difficulty of anticipating the price cycle, which is highly unpredictable in terms of the time, frequency, and amplitude of price spikes in competitive electricity markets (White 2006). Prices remain close to the variable cost of the technologies of the marginal producer until the hourly demand approaches the production capacity, at which time the prices increase dramatically to high levels, reflecting a scarcity rent (for instance, €1,000 per megawatt-hour (MWh) instead of €50/MWh in baseload hours). The net revenues (power price less fuel, operating and maintenance costs) during price spikes will contribute mainly to the capital cost amortization of the new equipment over a number of cycles. Consequently, for the developer of large and costly facilities such as nuclear equipment, the synchronicity of their commissioning with upcoming price spikes will have an important impact on the financial performance of the project, whereas long construction lead times make it impossible to anticipate upcoming price spikes.

The main problems with investing in capital-intensive equipment in a market environment are risk management and financing. There are high upfront construction costs but relatively low operating costs, while for gas-fired facilities the reverse is true. Nuclear investment combines huge fixed costs with political and regulatory risks. While the investment cost per kilowatt (kW) is three to four times higher than that for the smaller, more modular CCGT technology ($2,000/kW instead of $500/kW), the size of the equipment is two to five times larger (1,000–1,600 MW compared to 300–600 MW), which means that the investment unit is ten times larger. Three-billion-dollar to four-billion-dollar projects are a challenge for electric companies with market values of twenty billion dollars or less as their stock value may be affected by the long lead-time investment and significant construction risk. In a market environment and open political system, long construction times are also a disadvantage in themselves because they increase the risk that the market environment and political circumstances will change, making the investment uneconomic before the project is completed. In comparison, CCGT plants can be licensed and built much more quickly and benefit from standardization and series effects.

Investment in clean coal plants has a cost structure and a risk profile that are intermediate ($1,000–$1,500 per kW, $1 billion for a 700 MW plant, 50% of the cost per kWh due to fuel costs). Investment costs for clean coal plants will tend to be similar to those for nuclear facilities ($2,000 per kW) if they are developed with carbon-capture devices and CO_2 sequestration.

The climate policies that are currently being adopted are intended to give an economic advantage to non-carbon equipment (new nuclear projects, renewables, and eventually CCS coal plants) by internalizing the CO_2 externality, incorporating it into the cost of competing technologies (coal and gas). However, there is a large uncertainty in the long-term CO_2 price, resulting from the choice of a cap-and-trade instrument rather than a foreseeable long-term tax, and from the low predictability of long-term objectives. Whereas the CO_2 price was expected to give clean coal a definitive advantage over coal and gas plants, in fact, the risk in the CO_2 price reduces the incentive to invest in nuclear plants and clean coal plants in the near future.

Given its advantages in terms of both capital intensity and the possibility of hedging the risk, combined cycle gas turbine (CCGT) technology appears to be unduly favoured over other investment choices at the expense of more capital-intensive equipment, such as clean coal thermal plants, large RES-E equipment (hydraulic, and wind farms), and nuclear plants. The expected levelized costs for CCGT indicate a significant advantage in the future, under the different scenarios of increasing gas price evolution and CO_2 cost internalization. Risk hedging clearly favours CCGT since its cost structure is dominated by fuel costs and since there is a correlation between fuel prices and electricity prices because a gas generation facility tends to be the marginal unit on the hourly market during the major part of the year (see below). Gas importation risks from Russia, Algeria, and other developing countries are limited by long-term contracting and the credibility stakes for their governments in the trade relations with EU member states. Investments in technologies such as coal, hydro, nuclear, and wind power, which do not benefit from these risk correlations, are more risky for producers. Producers need to allocate part of their investment risks to suppliers or large consumers through vertical arrangements (long-term contracts at a fixed price, producer-consumer consortia, or vertical integration of supply and production).

However, in the de-integrated market model, which was and continues to be the reference point for electricity reforms, these arrangements are impeded by regulation for the sake of competition (to favour entrants) or undermined by the specific characteristics of competition in the wholesale and retail markets. For example, retailers confronted by the permanent risk of customers switching off if they do not follow the changes in the wholesale market price are reluctant to commit to long-term fixed price contracts with producers.

This situation does not favour the pursuit of climate policy goals in the electricity sector. Large-scale, capital-intensive non-carbon genera-

Table 4.2. Characteristics of Costs and Risks for Different Electricity Generation Technologies

Technology	Levelized cost CO_2 price of \$0-\$30/t CO_2	Capital size per unit (millions)	Lead time	Capital cost share	Fuel cost share	CO_2 cost	Fuel price risk
CCGT (400–600 MW)	\$5.6–\$6.7/kWh	Low (\$100–\$200)	Short	Low	High	Medium	High but correlated to electricity price
Coal (2 x 700 MW)	\$4.2–\$6.6/kWh	Large (€700–€1,000)	Long	High	Medium	High	Medium but low correlation with electricity price
Nuclear (1,500 MW)	\$5.5–\$6.7/kWh	Very large (€2,000–€3,000)	Long	Very high	Low	Nil	Low
Renewables (Wind farm/ 200 MW)	\$7–\$8/kWh	Medium (€300)	Medium	Very high	Nil	Nil	Nil

Sources: Deutsch and Moniz 2003; IEA/NEA 2005.

tion technologies (nuclear, very large wind farms, coal CCS, and new hydro) tend to magnify market risks owing to their capital intensity, indivisibility, long lead times, and long life cycles. Moreover, they are penalized by problems of acceptability (CO_2 sequestration, nuclear plant siting, nuclear waste management siting, hydro siting) and regulatory and political risks. Small-scale non-carbon technologies (for example, wind power, mini-hydro, wave) face some of the same problems (capital intensity) but also encounter specific market barriers because they do not benefit from scale effects on administrative costs and political risks.

Solutions Other Than the Decentralized Market Model

Electricity producers, wholesale buyers, and consumers could normally manage price volatility if they were able to develop the contractual arrangements necessary for efficiently allocating the risks among generators, intermediaries, and consumers. Market risks could be transferred from the producers to other stakeholders (suppliers, large consumers) through long-term contracts at entry cost and eventually through options contracts. However, the reality is more complex.

The Intrinsic Difficulty of Hedging Investment with Long-Term Contracts. The decentralized market model supposes that pure producers easily protect supplier and consumer interests by hedging their own risks. 'Consumers,' that is, suppliers and large consumers, compete to buy electricity by entering into bilateral forward contracts with different generators, including new entrants, and by managing their risks through portfolio strategies on the power exchanges. Downstream, suppliers have to manage a portfolio of different types of contracts with time spans and price formulas adapted to different clientele segments, with volume risks inherent in the ability of customers to switch suppliers. They are supposed to harmonize risk management between the sourcing portfolio and the sales contract portfolio.

However, in the real world, wholesale buyers do not want to reveal their hedging strategies, which involve long-term contracts with specialized generators at fixed prices to avoid opportunity costs when the market price drops below the contractual price, and to limit the risk of losing market share. In the more concrete terms of risk management for a retailer, risks are not manageable by developing a portfolio of long-term contracts with new generators, or even by acquiring shares or buying bonds from different specialized generators, all of which are possibilities in ideal electricity and financial markets (Roques, Newbery, and Nuttall 2006). Given this problem, pure producers fail because it is difficult to establish long-term contracts at a fixed price (or an indexed price) with creditworthy buyers for developing new equipment, since retailers are averse to making long-term commitments. In the de-integrated market model, a generator cannot rely on long-term contracts to hedge its new generation investment. This could be regarded as a market barrier because of the impossibility of securing long-term revenues for new generation equipment in the pure market model.

Possible Solutions. Overcoming the difficulty of finding non-reluctant and creditworthy contractors in the long term requires a two-pronged solution. Contracts involving large consumers require capacity development in joint ventures with horizontal arrangements between associations of large consumers and producers. Retailers must either have an existing base of loyal consumers attached to historic suppliers or a remaining franchise for the supply of households (Green 2004; Newbery 2002); otherwise they must adopt Joskow's approach, which uses last-resort supplier provisions to maintain a large segment of loyal consumers (Joskow 2007), allowing retailers to pass on the price differential of their wholesale market purchase contract through their retail prices. Under these scenarios, historic suppliers will retain a large fringe of inactive consumers to which they can transfer part of the sourcing risks because these consumers have no interest in switching.

Finally, although vertical integration has been opposed by regulators, there are advantages to investing in a mix of generation layers. The vertically integrated generator controls the risks associated with asymmetrical changes in profit margins at each stage under the effect of market price fluctuations; what is lost by one unit is recaptured by another. From the perspective of the integrated supplier, this also stabilizes and secures the terms of its wholesale purchases, even if it does not completely control volumetric and price risks in resale. The advantage becomes even clearer when vertical integration is organized within a historic supplier that benefits from a large segment of core consumers, to whom some of the project and market risks can be transferred.

This suggests that imperfect market reforms which have preserved the essentials of the vertical and horizontal integration of historic utilities (aside from the unbundling of transmission systems from vertical companies) might appear to be more socially efficient than is usually presumed by liberal reformers. They allow the development of capital-intensive equipment at lower capital cost, with less risk aversion, and within the project management capability of large-scale companies.

New Risks with Market-Based Policy Instruments

The next section examines problems arising from the choice of cap-and-trade instruments for non-carbon investment.

The Risk Inherent in a CO_2 Cap-and-Trade System

European governments and the European Commission have preferred to act on climate policy through a market instrument that fixes emission quantities and allows the exchange of permits to determine a carbon price rather than through imposing a carbon tax. There are two reasons: the political unacceptability of a tax and a strong belief in market virtues.[4] In 1995 a European eco-tax was promoted by the European Commission before the establishment of the Kyoto Protocol, but it failed because of opposition from a number of member states influenced by industrial groups that feared loss of competitiveness. Finally, before signing the Kyoto accord, the member states agreed on a binding strategy with a quantitative objective (–8% between 1990 and 2010), sharing the burden among them (Germany, –21%; United Kingdom, –12.5%; Spain, +15%; France, 0%). This was followed by the adoption of the cap-and-trade instrument for the large industry sector contained in the 2003 EU Emission Trading System (ETS) directive, implemented in 2005, and by the introduction of a carbon price. During the first three years, the spot market and future prices were very high (around €25–€30) until April 2006 and then retreated rapidly to very low levels when operators and speculators understood that too many permits had been issued for this period.

The use of a tax is clearly much more attractive to investors. With a cap-and-trade instrument, carbon prices are subject to market forces, and thus much of the price uncertainty is typical market uncertainty. Unauthorized banking between periods artificially increases price volatility, rendering short- and mid-term prices more chaotic. Moreover, the lack of long-term price history implies that there is no sound basis upon which to anticipate future prices. Although investors may be aware of the CO_2 policy risk, they cannot assess the risk of low CO_2 prices in the absence of meaningful historic trends for CO_2 prices. They are likely to

Box 1. Main Characteristics of the European Union's Emission Trading System during the First Two Periods (2005–2007 and 2008–2012)

Seven industrial sectors are regulated by the system: the energy and electricity industry (including the combined heat and power of self-producers' steam boilers, and coke ovens), oil refining, glass, ceramics, cement, pulp and paper, and steel production (aluminum and non-ferrous metals are excluded). The number of installations affected is around ten thousand, 65% belonging to the energy sector and 20% to the mineral industry.

The system is defined on a periodic basis: three years for the first period, and five years for all the others. The agents do not know what the target will be for subsequent periods.

Banking and borrowing of permits is not authorized between periods.

Allocation by grandfathering: the permits are free, and the allocation of permits to an industrial agent is determined according to its past emissions, decreased by a fixed reduction. There are two other rules: free allocations for new projects, and cancellation of initial allocations following plant closures. Entrants with new equipment must either buy permits on the market or benefit from special allocations from a reserve at no cost, in order to help competition.

According to the definition of national allocation plans (NAPs), member states are responsible for determining the total number of permits, based on the present emissions of each sector and on the overall objectives, which differ from one member state to another. The plans specify the target assignments by sector and the cap assignments for the installations covered in each sector. The European Commission has the power to control the definition of NAPs to limit overly generous allocations by governments under the influence of industrial companies.

In the case of non-compliance, a penalty must be paid at the end of each year, which will increase in the future (€40/tCO_2 for the first period, then €100/tCO_2).

Exchanges have been established in different countries and tend to be increasingly connected.

Figure 4.2. Volatility of the CO_2 Price during the First Three-Year Period (2005–2007) of the EU Emission Trading System, in €/tonne of CO_2

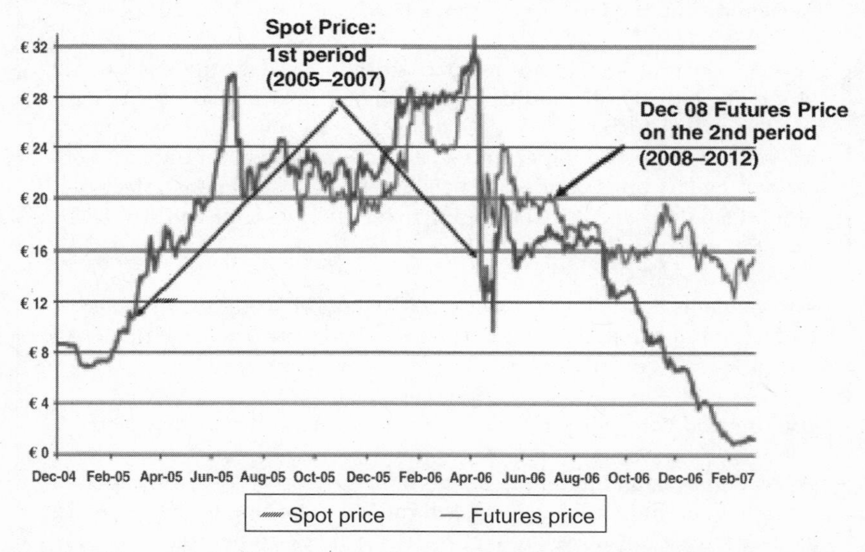

Sources: For spot price, Powernext Carbon; for December 2008 futures price, European Climate Exchange

underinvest in low-carbon technologies that could be competitive and profitable if a stable carbon price could be anticipated.

Over several decades, utility companies have seen how regulatory and policy choices determine investment outcomes. When assessing investment choices, they are guided mainly by current policy frameworks, such as the evolving electricity market regime and the European Union's ETS (Trochet, Bouttes, and Dassa 2008). Current prices, forward prices, and existing policies are the dominant drivers of investment choices, and only credible commitments to changes in these policies will affect decisions. In the absence of strong guidance, some utility companies might continue with traditional investment approaches, focusing mainly on diversification within coal and gas operations.

A cap-and-trade scheme is much more exposed to problems of learning and design than is a tax; it cannot predict the long-term carbon price, nor can it influence the competition between the main emitting technologies (CCGT, and coal plants) and non-carbon technologies (nuclear reactors, large hydroelectric power stations, large wind farms, and CCS) if it does not provide market players with carbon price foreseeability. Another source of uncertainty arises from the progressive, phased enforcement of the EU ETS, because the emissions cap cannot be predicted in the long term. Players cannot anticipate the evolutionary target or the carbon price. A third source of uncertainty is the interaction of carbon credits issued in developing countries and former Soviet bloc countries with the trading of European permits. Only credible long-term commitments to CO_2 reduction targets and rules will affect investment decisions in favour of 'clean' but more capital-intensive technologies (Ellerman et al. 2007; Ellerman and Joskow 2008).

Another critical issue is that the design of the ETS instrument has not favoured efficiency during the first period, owing to the nature of its main rules. Free allowances for new projects and cancellation of initial allowances after the closure of a facility are counter-incentives to factoring the implied carbon cost into decision making. This gives emitting technologies an added edge over least emitting technologies. Moreover, the renegotiation of quotas every five years can encourage producers to retain emitting equipment and invest in higher emission technologies to benefit from free allocations from the preceding periods. Finally, an analysis of investment in generation equipment in the European countries since 2005 shows that the ETS instrument has had no real influence on decision making; there have been few closures of old coal generation plants, some development of new coal plants in Germany and Italy, and no marked renaissance of nuclear investment even though companies are aware that the carbon price could have a trigger effect if it becomes stable in the future. The development of renewables depends upon specific instruments.

Finally, it is important to consider the redistributive effect resulting from the free allocation of permits in the electricity system, which has raised the issue of social acceptability. While they are receiving free allowances, electricity generators incorporate the carbon cost into their bids on the hourly electricity markets. Their actions are tantamount to trading in a carbon opportunity cost (that is, as if they would rather resell their permits on the CO_2 market than run their equipment). The result has been a supplementary increase in the electricity market price throughout the year because coal or gas generators are always marginal technologies on the hourly market. It has been estimated that during the first two years of the first period the major companies received a rent of around €5 billion (Pointcarbon 2008), without any real effect on investment in non-carbon generation technologies. In other sectors such as the cement industry, which are subjected to the pressure of international competition from countries without binding climate policies, the effect on price has been negligible because only the real cost of small CO_2 permits is passed through in the price of cement.

All these flaws have been widely debated in Europe. Some of them will be corrected in the directive of 2008, which defines the rules for the next period (2008–2012). In particular, it is envisaged that CO_2 permits will be allocated to electricity companies by auction in such a way that the CO_2 rent will be transferred to the public budget of the member states and not to the pockets of the electricity companies as a windfall profit. Clearly, a cap-and-trade instrument is much more complex to implement, and it tends to add new risks to the difficult job of investors. This suggests that the most straightforward way to limit the CO_2 price risk is through the design of climate policy. One approach, proposed by Ellerman and Joskow (2008) based on the lessons learned from the sulphur dioxide quota system in the United States, is a long-term CO_2 quota system, rather than a time-limited, phased quota system.

Different Strategies for Allocating CO_2 Risk Away from Producers in the Decentralized Model

Grubb and Newbery (2008) suggested limiting the price risk via a government long-term price declaration. This could be achieved in the detailed definition of the allowance mechanisms with the direct implementation of a floor on the carbon price and budget compensation for non-emitting producers.

A more radical approach would be a government guarantee of a minimum CO_2 price through long-term CO_2 contracts for each new carbon-free facility. This could be implemented through fixed-price contracts for avoided emissions, or option contracts auctioned before decisions to develop nuclear, CCS, or large renewables units (Ismer and Neuhoff 2006; Grubb and Newbery 2008). Contracts would be auctioned for emissions reductions over twenty to thirty years, and winning projects would achieve reductions at the lowest cost because they would have the lowest capital costs. The attractiveness of this approach to investors depends on the credibility of the government commitment to a long-term target price or, more concretely, on government willingness to respect the options contracts during their long time span (Helm, Hepburn and Mash 2003).

However, it is fair to say that such arrangements are not yet envisaged in any country, particularly in the United Kingdom, where the nuclear renaissance is politically accepted but only under the condition that investors will receive no public support.

Alternative Policy Instruments for the Promotion of Renewables' Electricity Units

With the focus on environmental issues, it is important to ask whether there is a need to promote renewables in situations where economic policies internalize the environmental externalities of polluting energy technologies and, in particular, the carbon cost. In principle, solutions exist that internalize the environmental effects of using fossil fuels and thus circumvent the need for specific instruments that favour renewables, such as a tax on CO_2 emissions or the cap-and-trade system. A criticism of specific policies favouring renewables is that the costs of avoiding CO_2 emissions (for example, \$100–\$150/t CO_2) could dramatically exceed the estimated social damage from CO_2 emissions (\$20–\$30/tCO_2) (see, for instance, Newbery 2003; Fischer and Newell 2008).

However, there is a fundamental difficulty in imposing a sufficiently high CO_2 tax (or stringent CO_2 quota) that would encourage the replacement of fossil fuels with renewables and technological progress: there is no guarantee that even a high price for CO_2 emissions will lead to more extensive substitution of new clean technologies for fossil fuels than will the policies directly supporting clean technologies (see, for instance, the literature reviewed in Jaffe, Newell, and Stavins 2002). The two main reasons are (1) regulatory uncertainty over the price of CO_2 (which, as mentioned above in relation to the EU ETS, could follow when the period during which the quota applies is too short) and (2) entry barriers for renewable technologies.

Accordingly, the European Union decided to act on the deployment of RES-E before establishing climate policies and the ETS covering the electricity industry. It issued a first directive in 2000, which imposed voluntary objectives on member states, requiring an average of 22% of RES-E from large hydro by 2010, up from 14% in 1997 (this means 8% from small RES-E), with some differentiation between member states based on RES potential (European Commission 2001). While the more recent EU objectives defined in January 2008 for the new directive in the so-called third energy-and-climate package are not more ambitious (22% of the electricity system, including large hydro, to be reached by 2020), the objectives will be binding on member states because the 2010 objective has not been achieved (4% from small RES-E, instead of 8%).[5]

There is a long-standing debate within the European Union between countries adopting the tradable green certificate (TGC) quotas supported by the European Commission and the most market-oriented instruments and the proponents of feed-in tariffs (FITs) based on an RES-E purchase obligation at an administered fixed price for each new unit, guaranteed over the long term (Finon and Perez 2007). See table 4.3 and

Table 4.3. Differences in the Choice of Instruments within the European Union

Countries with feed-in tariffs	Countries with tradable green certificates / quotas
Austria	United Kingdom
Denmark	Italy
Finland	Sweden
France	Belgium
Germany	
Greece	
Ireland	
The Netherlands	
Portugal	
Spain	

Note: Among the ten new member states, only Poland has adopted tradable green certificates / quotas.

appendix 4.1 for the detailed design of the instruments chosen by the EU-15 member states. The proponents of the TGC system argue that its quality market incentive will increase efficiency and that FITs have higher costs for consumers. However, both TGCs and FITs are reasonable in light of recent experience.[6]

The next section describes the characteristics of the two instruments and then discusses the lessons that can be drawn from the experience of some member states in promoting RES-E via these two instruments.

Salient Features of Feed-In Tariffs and Tradable Green Certificate Systems

A feed-in tariff is an obligation to purchase electricity based on renewable energy at a fixed (fairly high) price. Both the obligation and the price guarantee extend over a long period, for example, eight years in Spain, fifteen years in France, and twenty years in Germany. The purchase obligation is restricted to distributors-suppliers in the service area and applies to all new renewable power generation units. To promote the development of a diverse set of renewable technologies, feed-in tariffs are technology specific and differ across technologies. They reflect the generating costs of a typical renewable electricity unit (including some risk premium) and are not set on the basis of the avoided generating cost of the distributor-supplier (as was the case for RES-E and CHP units before 1995). Unless the supply curves for renewable electricity are known, the quantity of renewable electricity production resulting from setting feed-in tariffs is not known ex ante.

The recovery of the extra cost of renewable electricity incurred by mandated buyers can be accomplished in three ways: an increase in the price of every kilowatt-hour sold by the distributors, subject to a purchase obligation when such distributors have a legal monopoly; compensation among competing distributors-suppliers, given that they are obliged, irrespective of their own sales, to buy all the renewable electricity produced in the area of their distribution networks; or reimbursements financed by a tax on all electricity transmitted via the national grid. In the latter case, the extra cost of renewable electricity is paid by all electricity consumers. An alternative or complement to passing on the extra cost to electricity consumers is budgetary support for mandated buyers. Budgetary support could also be given to producers of renewable electricity to limit the level of feed-in tariffs and, thus, the cost to consumers; this could be achieved either through an eco-tax and/or a value-added-tax exemption, as in the Netherlands and Denmark, or tax credits for renewable electricity production, as in the United States.

Tradable green certificate systems designate economic agents that will be subjected to a rising renewable, or green, electricity quota (typically electricity suppliers or distributors/retailers), and identify eligible technologies and installations, generally including only new installations and possibly excluding new large hydro plants and waste incineration facilities. Since the different certificate markets are very small, no banding of technologies is considered. This is a serious limitation of the instrument, requiring that additional support be provided for the other technologies.

Designated agents, referred to as suppliers, can fulfil their quotas (expressed as a percentage of each supplier's annual electricity sales, rising over time) in different ways. They can produce renewable electricity, purchase it under long-term contracts from specialized producers, or purchase green certificates from suppliers who exceed their quotas or from specialized producers who choose to sell part of their renewable electricity in the market rather than directly, under long-term contracts.

The quota is complemented by a penalty to be paid in the case of non-fulfilment. This penalty could be seen as a price cap rather than as a threat to force suppliers to meet their quotas. Rather than fulfilling his quota, a supplier may opt to pay the 'buy-out price' (the UK term) for not meeting it, which could, in extreme cases, represent the full quota. In essence, the buy-out price puts a ceiling on the cost of renewable certificates. The last characteristic of the TGC design is the reallocation of the revenue from the penalties to agents who have strictly respected their quotas, which provides an added incentive to abide by them.

Lessons from Experience

The experience of EU member states in promoting renewable electricity production is now sufficiently well documented that lessons can be drawn on how various instruments have worked in practice (Finon and Menanteau 2003; Finon and Perez 2007). Insights can be derived from the experience gained through designing and applying various policy instruments and from an analysis of what they have achieved in terms of meeting policy objectives.

The first lesson is that the influence of a particular instrument cannot be isolated from other factors that foster or hinder the development of a country's renewable electricity resources.[7] Specifically, the success of an instrument depends not only on the level of support it provides but also on the protection it offers from the risks encountered in the planning and siting procedures, and on the rules that govern the recovery of both the balancing costs for intermittent production and the cost of connecting renewable power plants to the network. For example, in 2000, France adopted feed-in tariffs as generous and predictable as those in Germany, but investment in renewable generating capacity and its performance fell far short of what was achieved in Germany. In 2006, installed wind energy capacity amounted to only around 1,700 MW in France, in contrast to 20,000 MW in Germany.[8]

The second lesson is that differences in the risks associated with support for renewable electricity largely explain why some European countries were more successful than others in increasing the share of renewable electricity. Before the difference in results can be discussed, it is necessary to explain that TGC systems incorporate far more uncertainties than do feed-in tariffs, resulting in risk aversion and higher risk premiums which suppliers, producers, and financiers must take into account when embarking on renewable electricity projects; this raises the cost of such projects. Consider the revenue characteristics of a renewable electricity project under each of the two instruments. In the case of feed-in tariffs, revenues are fairly certain because there is a guaranteed price at which production can be fed into the network. In the case of TGC systems, revenues depend on the uncertain market price of electricity and the uncertain price of green certificates; electricity price risk contains green certificate price risk. Moreover, when the production of renewable electricity is difficult to schedule, as it is for wind energy, the electricity price risk is exacerbated by uncertainties arising from the balancing costs, which, in TGC systems, are borne entirely by the producers of renewable electricity. In TGC systems, the generation of renewable electricity must observe all electricity market rules, including those pertaining to the market balancing mechanism, which ensures the reliability of the whole power system (Mitchell, Bauknecht, and Connor 2004). In contrast, under feed-in tariffs, renewable power plants do not need to supply a specified load profile, and the balancing costs fall on obligated suppliers.

Revenue risk also arises from uncertainty as to how the quota will increase over time and, in particular, uncertainty as to the level beyond which it will cease to be raised. When the quota approaches its limit, investment in additional renewable electricity generating capacity may create an oversupply of green certificates, which will then drop in price. This increases the risk involved in renewable energy projects in a TGC system and, thus, their costs.

The differences between instruments can also be discussed in the context of onshore wind installations, so far the most successful, emergent, renewable electricity technology. Table 4.4 shows that there are large differences in wind power capacity and in growth of capacity among six selected countries. We use a meaningful indicator: the per capita installed capacity (in watts per capita) in wind power facilities installed between 2001 and 2007. In Austria, Germany, and Spain (countries that offer feed-in tariffs) and in Belgium, Italy, Sweden, and the United Kingdom (countries that offer TGC systems), we can assess the efficiency of the two support mechanisms for renewable electricity. If the level of installed capacity during a period is used as a measure of the environmental effectiveness of the underlying policy, then the countries with feed-in tariffs performed much better than did the countries with TGC systems.

Austria, Germany, and Spain have applied feed-in tariffs from 1998 to the present. Combined with low administrative barriers, this stimulated a strong, continuous growth in wind energy. In contrast, in the Belgium, United Kingdom, Italy, and Sweden, the transition from a tendering system or feed-in tariffs to a TGC system created substantial

Table 4.4. Comparison of Wind-Power Capacity of Selected Member States

	Instruments	Capacity end 2000 (MW)	Capacity end 2007 (MW)	Capacity increase 2001–2007 (MW)	Increase per capita (W/capita)
Germany	FIT*	6,113	22,247	16,364	200
Spain	FIT(variant premium)	2,235	15,145	12,910	290
Italy	TGC**/quota	427	2,726	2,299	38
UK	TGC/quota	406	2,388	1,982	32
Austria	FIT	77	981	904	110
Sweden	TGC/quota	231	653	419	45
Belgium	TGC/quota	194	287	93	9

*Feed-in tariff
**Tradable green certificate
Source: European Wind Energy Association Statistics (English version)

uncertainty. The indicator of installed wind-power capacity per capita shown in table 4.4 (last column) provides strong evidence for the hypothesis that TGC systems do not create an environment sufficiently secure to foster investment in renewable electricity generation.

It is often presumed that feed-in tariffs offer more generous support for renewable electricity than do TGC systems and that this explains their greater environmental effectiveness. However, the predictability of feed-in tariff support is an important factor because, from the perspective of potential producers and investors, TGC systems involve considerable risks.

One consequence is that the revenue required to induce investment in renewables is higher under TGC systems than with feed-in tariffs. Empirical support for this hypothesis comes from Butler and Neuhoff (2004), which showed that the remuneration for wind energy is higher under the UK TGC system than with the German feed-in tariffs, which are often portrayed as excessively generous. More precisely, their study showed that the remuneration for wind energy ranges from €77/MWh to €100/MWh in the British mechanism, compared to €70/MWh with Germany's feed-in tariffs. Similar evidence was provided by Ragwitz et al. (2006), which estimated expected revenues for new producers of onshore wind energy.

Figure 4.3 shows that expected revenues are much higher in countries using TGC systems than in those relying on feed-in tariffs.[9] Sweden's success results from the specificity of its TGC system, which was not

Figure 4.3. Expected Revenues (€/MWh) for Onshore Wind Energy in Selected Member-States: Feed-In Tariffs vs the Tradable Green Certificate (TGC) System

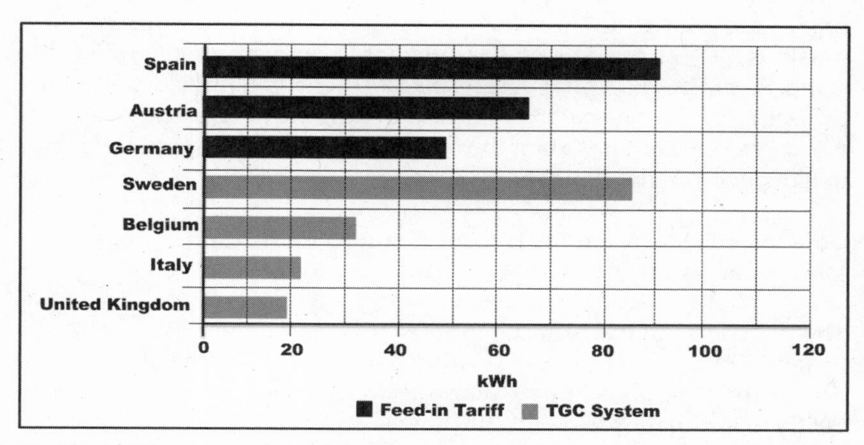

Note: Estimates reflect levelized expected revenues.
Source: Ragwitz et al. (2006)

adopted until 2003 and replaced a system that offered large tax credits and investment subsidies; Sweden's statistics include existing installations and cheap technologies using biomass (for example, combined heat and power, and co-firing) in the portfolio of eligible technologies. It is reasonable to conclude that, in practice, feed-in tariffs do not provide exceptionally high revenues to producers and that the reliability and predictability of the policy and investment environment are crucial to the successful development of the market for renewable electricity.

The third lesson to be drawn on how various instruments have worked in practice is that governments must offer investment support (soft loans, tax allowances, and subsidies) to complement TGC systems. The need for such schemes seems to be greater in the case of TGC systems, whose objective is to foster not only the most technologically and commercially advanced renewable option but also to encourage options that lag behind.

In conclusion, in light of recent experience with competing instruments adopted to promote renewable electricity, it is not surprising that strong supporters of TGC systems have become more cautious, as evidenced by the evolving position of the European Commission (European Commission Directorate-General Energy and Transport 2005, 2008). Policymakers are increasingly aware, as they should be, of the complexity of the innovation process driving renewable electricity technologies. Once they have chosen an instrument, it is incumbent upon them to signal clearly that the support mechanism will remain in place long enough to ensure an acceptable return to producers. They must also assess the risk profile inherent to an instrument and understand its effect on capital cost, especially when combined with other project risks, in particular political and administrative risks. In other words, the cost of mitigating the risk aversion of developers and investors depends on the choice of instrument.

Conclusion: The Need for Hybrid Regimes for Sustainable Development

The electricity industry in Ontario is still under the influence of the former regulatory regime, exhibiting features such as the central position of the publicly owned Ontario Power Generation, the freezing of tariffs, coordination by Ontario Power Authority auctions for long-term contracts with entrants, the programming of renewables development and demand side management savings, and the phase-out of coal generation by 2014 to meet environmental objectives. These could facilitate planning, but they do not provide incentives for efficiency through market pressures or through the internalization of environmental costs, which gives realistic price signals to electricity producers and consumers.

Ontario is contemplating market liberalization in concert with an ambitious policy of CO_2 emissions reduction, which will diminish the environmental footprint of the electricity system. However, the European experience suggests that a better strategy is to adopt a hybrid model that

Table 4.5. Comparison of Electricity Regimes and Environmental Policy Design Based on Long-Term Effectiveness

	Market model	'Cost of service' regulatory model
Effectiveness	Risk aversion and slow development of non-carbon technology	Effective programming
Effects on investment risks and costs	Market risks and regulatory risks	Risk of uneconomic decisions

Table 4.6. Comparison of Market, Regulated Utility, and Hybrid Regimes

	Market model	'Cost of service' regulatory model	Hybrid model
Industrial structure	De-integrated firms	Vertical integration and supply monopoly	Variety of firms and vertical arrangements; oligopoly
Short-term coordination	Hourly market and short-term contracts	Engineer's merit order dispatching	Vertical integration; dispatchable contractual producer and some spot transactions
Generation capacity development	Decision of decentralized players	Investment programming, regulatory authorization, and prudency reviews	Coordination by programming and auctioning; long-term fixed price contracts; industrial and load-serving entity consortia
Type of environmental policy instrument	Market-based instrument. Market-oriented (tax)	Emission standards, technology substitu-tion programming	Compatibility with the two types of instruments
RES-E development	Market-based (exchangeable quotas)	Purchase obligation on utility and feed-in tariff	Preferably feed-in tariffs
Electricity efficiency development	Price signals are assumed efficient	Demand side management and electricity efficiency schemes	Energy efficiency obligation on suppliers

incorporates market design and vertical industrial structures. These elements do not deter investment in capital-intensive equipment, because they avoid the risks created by reliance on the short-term market for the prices of electricity and environmental goods (see table 4.6).

It would be helpful to define a hybrid model. Imperfect power industry liberalization must preserve in part the former mode of coordination of the regulated monopoly. The main function of the electricity market should be to make short-term adjustments, but efficiency incentives should be created through competition, by encouraging an entry thread and facilitating competitive pressure on imports through the integration of regional markets. To orient market players' choices towards clean and non-carbon technologies, policy instruments in a hybrid regime must require price signals for environmental goods that are foreseeable and stable over the long term. This can be achieved with an increasing carbon tax in the climate policy and a feed-in-tariff for renewables. The Ontario electricity regime seems to be open to such a vision.

Appendix

Table 4.7. Promotion Strategies for RES-E in the EU-15 Countries in 2007

		RES-E TECHNOLOGIES CONSIDERED IN EU COUNTRIES		
	Major strategy	Large hydro	Small hydro; 'New' RES (Wind on- and offshore, photovoltaic, solar thermal electricity, biomass, biogas, landfill gas, sewage gas, geothermal)	Municipal solid waste
Austria	FITs	No	Renewable energy act (Ökostromgesetz) 2003. Technology-specific FITs guaranteed for 13 years for plants that get all permissions between 1 January 2003 and 31 December 2004 and, hence, start operation by the end of 2006. Investment subsidies mainly on regional level. No decision yet on follow-up support after 2004.	FITs for waste with a high share of biomass
Denmark	FITs and premiums, net metering for photovoltaic	No	Act on payment for green electricity (Act 478): Fixed premium prices instead of former high FITs for onshore wind. Tendering for offshore wind. Biomass and biogas receive FITs of €80/MWh for the first 10 years, €54/MWh for the next 10 years. Net metering for PV used for individual houses	Subsidies available for CHP plants using waste.
France	FITs	No	Up to mid-2007, FITs for RES-E plant < 12 MW (wind plants are not subject to the capacity limit) guaranteed for 15 years (PV and Hydro for 20 years). Tenders for plant >12 MW. After mid-2007, no limitation of capacity for FITs, provided that the facilities are in specific zones decided by local communities and regional administrations (Energy act of July 2005). FITs in more detail: biomass, €49–61/MWh; biogas and methanization, €75–90/MWh, including premium for energy efficiency up to €120/MW, including premium for 'methanization' up to €140/MWh; geothermal, €76–79/MWh; PV, €300/MWh (20 years) including premium (for 'integration in buildings') up to €550/MWh; sewage and landfill gas, €45–60/MWh; onshore wind,[a] €28–82/MWh; offshore wind,[b] €30–130/MWh; hydro, €54.9–61/MWh.	FIT: €45–50/MWh
Germany	FITs	Only refurbishment	Novelty in German renewable energy act in 2004: FITs guaranteed for 20 years. FITs for new installations (2006) in more detail: hydro, €66.5–96.7/MWh (30 years); wind,[c] €52.8–83.6/MWh; biomass and biogas, €81.5–171.6/MWh; landfill, sewage and landfill gas, €64.5–74.4/MWh; PV, €406–568/MWh; geothermal, €71.6–150/MWh.	No
Ireland	FIT	No	FITs are granted for 15 years. Tariff level (2006): wind: €57–59/MWh; landfill gas: €70/MWh; other biomass, €72/MWh; small hydro, €72/MWh.	No
Netherlands	No support system	No	FIT scheme was abolished in summer 2006 since the government expects to fulfil the 2010 target set by the European Commission without further financial support, and RES-E support costs were higher than expected.	No
Portugal	FITs + investment subsidies	No	FITs (Decree Law 33-A/2005) and investment subsidies of roughly 40% (measure 2.5 [MAPE] within program for economic activities) for wind, PV, biomass, small hydro, and wave. Average FITs in 2006: wind,[d] €74/MWh; wave: n.a.; PV,[e] €310–450/MWh; small hydro, €75/MWh.	FIT for urban waste: €75/MWh

[a]Stepped FIT: €82/MWh for the first 10 years of operation and then €28–€82/MWh for the next 5 years, depending on the quality of site.
[b]Stepped FIT: €130 /MWh for the first 10 years of operation and then €30–€130/MWh for the next 10 years, depending on the quality of site.
[c]Stepped FIT: In case of onshore wind, €83.6/MWh for the first 5 years of operation and then €52.8–€83.6/MWh depending on the quality of site.
[d]Stepped FIT depending on the quality of the site.
[e]Depending on the size: <5kW, €420/MWh; or >5kW, €224/MWh.

Table 4.7. (*Concluded*)

<table>
<tr><th></th><th></th><th></th><th colspan="3">RES-E TECHNOLOGIES CONSIDERED IN EU COUNTRIES</th></tr>
<tr><th></th><th>Major strategy</th><th>Large hydro</th><th>Small hydro</th><th>'New' RES (Wind on- and offshore, photovoltaic, solar thermal electricity, biomass, biogas, landfill gas, sewage gas, geothermal)</th><th>Municipal solid waste</th></tr>
<tr><td>Spain</td><td>FITs or fixed premiums</td><td>Depend ing on the plant size</td><td colspan="2">FITs (Royal decree 436/2004): RES-E producers have the right to opt for a fixed FIT or for a premium tariff.[f] Both are adjusted by the government according to the variation in the average electricity sale price. In more detail (2006): wind, biomass, small hydro (<25 MW), geothermal, €68.9/MWh (fixed) and €38.3/MWh (premium); solar thermal and PV:[g] €229.8–440.4 /MWh.

Agricultural and forest residues: €61.3/MWh (fixed) and €30.6/MWh (premium). Moreover, soft loans and tax incentives (according to 'Plan de Fomento de las Energías Renovables') and investment subsidies on regional level.</td><td>FIT: €53.6/MWh (fixed) or €23/MWh (premium)</td></tr>
<tr><td colspan="6">THE TGC/QUOTA SYSTEM</td></tr>
<tr><td>Italy</td><td>TGC/Quota + FITs for PV</td><td>Included in TGC system</td><td colspan="2">Quota obligation (TGC system) on electricity suppliers: 3.05% target (2006), increasing yearly 0.35% until 2008; TGC issued for all (new) RES-E (incl: large hydro and municipal solid waste) – with rolling redemption; no penalty; €12,528/MWh (2006) to purchase TGCs from the grid operator, but market distortions appear; feed-in tariff for PV and local investment subsidies from regional administrations.</td><td>Included in TGC system</td></tr>
<tr><td>Belgium</td><td>TGC/Quota + guaranteed electricity purchase</td><td>No</td><td colspan="2">Federal: The Royal Decree of 10 July 2002 (in effect from 1 July 2003) sets minimum prices (i.e., FITs) for RES-E. On regional level promotion activities include
– Wallonia: quota-based TGC system for electricity suppliers, increasing from 3% in 2003 to 12% in 2010.
– Flanders: quota-based TGC system on electricity suppliers, increasing from 3% (no MSW) in 2004 up to 6% in 2010.
– Brussels region: no support scheme yet implemented.</td><td>Included in TGC system (except Flanders)</td></tr>
<tr><td>Sweden</td><td>TGC/Quota</td><td>No</td><td colspan="2">Quota obligation (TGC system) on consumers: increasing from 7.4% in 2003 up to 16.9% in 2010. For wind, investment subsidies of 15% and additional small premium FITs ('Environmental Bonus')[h] are available.</td><td>No</td></tr>
<tr><td>United Kingdom</td><td>TGC/Quota</td><td>No</td><td colspan="2">Quota obligation (TGC system) for all RES-E: increasing from 3% in 2003 to 10.4% by 2010; buyout price is set at £32.33/MWh for 2005/2006.
In addition to the TGC system, eligible RES-E are exempt from the Climate Change Levy certified by levy exemption certificates (LECs), which cannot be traded separately from physical electricity. The current levy rate is £4.3/MWh.
Investment grants in the frame of different programs (e.g., Clear Skies Scheme, DTI's Offshore Wind Capital Grant Scheme, the Energy Crops Scheme, Major PV Demo Program, and the Scottish Community Renewable Initiative).</td><td>No</td></tr>
</table>

Source: Haas et al. (2008)

[f]In the case of a premium tariff, RES-E generators earn in addition to the (compared to fixed rate, lower) premium tariff the revenues from the selling of their electricity on the power market.
[g]In the case of PV, the expressed premium tariff refers to plants >100 kW. For small-scale plants (<100 kW), only a fixed FIT is applied.
[h]Decreasing gradually down to zero in 2009.

NOTES

1 With the help of different 'directives' to be transcribed in the national laws of member states, the European Union will influence national energy policies, the technological evolution of electricity systems, and the rationalization of energy use.

2 Jurisdictions can choose to create mandatory markets to optimize dispatching as in regional U.S. markets (for example, Pennsylvania-New Jersey-Maryland and New York).

3 When institutional economists consider telecommunications reforms (Levy and Spiller 1996), they must take into account the institutional and socio-political environment of an industry. The feasibility and the credibility of a competition-based reform model are different in different institutional environments. The feasibility of a reform (that is, the ability to implement it despite the influence of the losers on the decision-making process through access to the courts or the electoral rules) and the credibility of a reform (that is, the guarantee that the reform will not deviate from its course or prove abortive, and the predictability of market rules) rely on the compatibility of the characteristics of a particular reform model and the characteristics of the institutional environment of each country.

4 This choice was also supported on the basis of academic considerations of the merits of price instruments and quantity instruments (quota in a cap-and-trade), taking into account ignorance of and uncertainty in the cost curves for damage and reduction costs.

Weitzman (1974) makes two points: A *price* mechanism is superior when the curve of marginal environmental damage is relatively flat; that is, in the case of quantitative variability, the marginal cost of damage shows little variation or uncertainty. A *quantity* mechanism is superior when the curve of the marginal cost of reduction is relatively flat; that is, when for the quantities emitted, the marginal cost of reduction shows little variation or uncertainty, which appears to be the case for the CO_2 problem.

Pizer (2002) adds that the social efficiency of a quantity instrument in a context of uncertainty can be improved by imposing a price cap (a safety valve) and a price floor. In the first case, agents prefer to pay a buy-out price (which could also be called a penalty) rather than respect their quotas if their marginal reduction cost is high; in the second case, when the market price of the permit decreases below a threshold, the government would pay the difference. Both measures would give the carbon price a certain foreseeability.

5 In fact, the binding 20% objective also refers to the total of renewable

energies (biofuel, solar, geothermal electricity), but the detailed objectives in the official scenario amount to 22% for the RES-E.

6 Finally, after intensive debates from 1998 to 2000 before the adoption of the 2000 RES-E directive and again between 2005 and 2008 before the adoption of the 2008 RES-E directive, it was decided that the two directives should allow member states to choose their instruments without seeking harmonization throughout the TGC system.

7 There is extensive literature about the causal links between the diffusion of renewable electricity and the variation in the design and strength of policy instruments. Examples include Reiche 2005, Meyer 2003, and van Dijk et al. 2003.

8 Key obstacles to developing renewable electricity generation in France are fragmented planning procedures and problems concerning local acceptability. Effective planning procedures and network integration rules can help reduce project costs and risks, so they must be an integral part of a successful renewable energy policy. However, as is the case with renewable energy in general, the political backing they receive is no stronger than that for the underlying renewable technology. Moreover, they mirror social preferences for global environmental protection and energy security on the one hand and local environmental concerns on the other.

9 Nonetheless, the difference will probably decrease as institutional experience with the relatively new TGC instruments accumulates, but even if it does decrease, this would not reduce the risk premium associated with the production of renewable electricity under TGC systems.

REFERENCES

Butler, L., and K. Neuhoff. 2004. 'Comparison of feed in tariff, quota, and auction as mechanisms to support wind power development.' Working paper, Department of Applied Economics, University of Cambridge.

Davies, C. 2008. 'Draft report on the proposal for a directive of the European Parliament and of the Council on the geological storage of carbon dioxide and amending Council Directives.' European Parliament, Committee on the Environment, Public Health and Food Safety. June.

Deutsch, J.M., and C. Moniz, eds. 2003. *The future of nuclear power: An interdisciplinary MIT study.* Cambridge, MA: Massachusetts Institute of Technology.

De Vries L., and K. Neuhoff. 2004. 'Investment incentives for investment in electricity generation.' Working paper in Economics, University of Cambridge.

Dijk, A.L. van, L.W.M. Beurskens, M.G. Boots, M.B.T. Kaal, T.J. de Lange, E.J.W. van Sambeek, and M.A. Uyterlinde. 2003. 'Renewable energy policies and market developments.' Energy Research Centre of the Netherlands, Policy Studies Report ECN-C-99-072.

Ellerman, A. Denny, Barbara K. Buchner, and Carlo Carraro, eds. 2007. *Allocation in the European Emissions Trading Scheme: Rights, rents and fairness.* Cambridge and New York: Cambridge University Press.

Ellerman D., and P. Joskow. 2008. 'The European Union's Emissions Trading System in perspective.' Report to the Pew Center, Washington, DC.

European Commission. 2001. 'Directive of the European Parliament and of the Council on the promotion of electricity from renewable energy sources in the internal electricity market.' Com(2001) 27 9 final, Brussels.

– 2003. 'Directive concerning common rules for the internal market in electricity.' 2003/54/EC, Brussels.

– 2005a. 'The support of electricity from renewable energy sources.' Communication from the Commission, Com (2005) 627, Brussels, 7.12.2005.

– 2005b. 'Annual report on the implementation of the gas and electricity internal market.' Report from the Commission, Sec(2004) 1720, Brussels, 5.1.2005.

– 2006. 'Directive du Parlement européen et du Conseil relative à l'efficacité énergétique dans les utilisations finales et aux services énergétiques.' Directive 2006/32/CE, Bruxelles, avril 2006.

– 2008. 'Proposal for a directive of the European Parliament and of the council on the promotion of the use of energy from renewable sources.' COM(2008) 19 final, 2008/0016 (COD).

European Commission Directorate-General Energy and Transport. 2005. *Implementing the internal energy market: Fourth benchmarking report.* Brussels.

– 2008. 'European energy and transport scenario to 2030.' European Commission report, Brussels.

Finon, D. 2008. 'Investment risk allocation in restructured electricity markets: The need of vertical arrangements.' *OPEC Energy Review* 32, no. 2 (June 2008): 150–83. See also Working Paper LARSEN no. 11, www.gis-larsen.org.

Finon, D., and P. Menanteau. 2003. 'The static and dynamic efficiency of instruments of promotion of renewables.' *Energy Studies Review* 12, no. 1:53–83.

Finon, D., and Y. Perez. 2006. 'The social efficiency of instruments for the promotion of renewables in the electricity industry: A transaction cost perspective.' *Ecological Economics*, no. 62:77–92.

Finon, D., and F. Roques. 2008. 'Contractual and financing arrangements for new nuclear investment in liberalised markets: Which efficient combination?' *Competition and Regulation in Network Industries Journal* 9, no. 3 (September): 247–82. See also Working Paper LARSEN no. 14, www.gis-larsen.org.

Fischer, Carolyn, and Richard G. Newell. 2008. 'Environmental and technology policies for climate mitigation.' *Journal of Environmental Economics and Management* 55, no. 2 (March): 142–62.

Green, R. 2004. 'Retail competition and electricity contracts.' CMI working paper 33, Cambridge University.

Green, R., and T. McDaniel. 1998. 'Competition in electricity supply: Will "1998" be worth it?' *Fiscal Studies* 19, no. 3.

Grubb, M., and D. Newbery. 2008. 'Pricing carbon for electricity generation: National and international dimensions.' In *Delivering a low carbon electricity system: Technology, economics and policy,* ed. M. Gubb, T. Jamasb, and M.G. Pollit. Cambridge: Cambridge University Press.

Haas, R., D. Finon, A. Held, N.I. Meyer, and R. Wiser. 2008. 'Promoting renewable energy sources for electricity generation: A historical overview.' In *Competitive Electricity Markets*, ed. F.P. Shioshansi. London: Elsevier.

Helm, D., C. Hepburn, and R. Mash. 2003. 'Credible carbon policy.' *Oxford Review of Economic Policy* 19, no. 3: 438–450.

Huber, C., R. Haas, and T. Faber. 2001. *'Action plan for a green European electricity market.' Compiled within the project 'ElGreen.'* International Energy Workshop (IEW).

International Energy Agency. 2003. *Power generation investment in electricity markets*. Paris: OECD.

– 2005. *Lessons from liberalised electricity markets.* Paris: OECD.

– 2007a. *Tackling investment challenges in power generation.* Paris: OECD.

– 2007b. *Climate policy uncertainty and investment risk.* Paris: OECD.

– 2007c. *Energy security and climate policy: Assessing interactions*. Paris: OECD.

International Energy Agency / Nuclear Energy Agency. 2005. *Comparison of electricity generation costs*. Paris: IEA-OECD.

Ismer R., and K. Neuhoff. 2006. 'Commitments through financial options: A way to facilitate compliance with climate change obligations.' EPRG0625.

Jaffe, A.B., R. Newell, and R.N. Stavins. 2002. 'Environmental policy and technological change.' *Environmental and Resource Economics* (22:1-2): 41–69.

Joskow, P. 2007. 'Electricity markets and investment in new generating capacity.' In *The new energy paradigm*, ed. by D. Helm. Oxford: Oxford University Press.

Levy, B., and P Spiller. 1996. *Regulations, institutions and commitment: Comparative studies of telecommunications.* Cambridge: Cambridge University Press.

Meyer, N. 2003. 'European schemes for promoting renewables in liberalised markets.' *Energy Policy*, no. 31:665–76.

Mitchell, C., E. Bauknecht, and P.M. Connor. 2004. 'Effectiveness through risk reduction: A comparison of the renewable obligation in England and Wales and the feed-in system in Germany.' *Energy Policy*, no. 32:1935–47.

Newbery, D. 2002. 'Regulatory challenges to European electricity liberalisation.' CMI working paper.

– 2003. 'Government intervention in energy markets.' British Institute of Energy Economics' conference, 25 September 2003.

Ontario Power Authority. 2007. *Overview of the IPSP.* August 2007.

Pizer, William A. 2002. 'Combining price and quantity controls to mitigate global climate change.' *Journal of Public Economics* 85, no. 3: 409–34.

Pointcarbon. 2008. 'EU ETS Phase II: The potential and scale of windfall profits in the power sector.' Pointcarbon Advisory Services, 29 pp.

Ragwitz, M., A. Held, F. Sensfuss, C. Huber, G. Resch, T. Faber, R. Haas, R. Coenraads, A. Morotz, S.G. Jensen, P.E. Morthorst, I. Konstantinaviciute, and B. Heyder. 2006. 'Assessment and optimisation of renewable support schemes in the European electricity market.' OPTRES interim report.

Reiche, D. 2005. *Handbook of renewable energies in the European Union: Case studies of the EU-15 states*. Berlin: Peter Lang Publishers.

Roques, F., D. Newbery, and W. Nuttall. 2006. 'Valuing portfolio diversification for a utility: Application to a nuclear power investment when fuel, electricity, and carbon prices are uncertain.' *The Energy Journal* 27, no. 4.

Sijm, J., K. Neuhoff, and Y. Chen. 2006. 'CO_2 cost passthrough and windfall profits in the power sector.' *Climate policy*, no. 6:49–52.

Trebilcok. M.J., and R. Hrab. 2006. 'Electricity reforms in Canada.' In *Electricity market reform: An international perspective*, ed. F.P. Shioshansi and W. Pfannenberger, Elsevier Global Energy Policy and Economic Series, xxx. Kidlington, UK: Elsevier.

Trochet, J.M., J.P. Bouttes, and F. Dassa. 2008. 'Assessment of EU CO_2 regulation.' In *Abatement of CO_2 emissions in the European Union*, ed. J. Lesourne and J.H. Keppler. Paris: IFRI Publication (Collection Les etudes).

Weitzman, M.L. 1974. 'Prices vs. quantities.' *The Review of Economic Studies* 41, no. 4:477–91.

White, A. 2006. *Financing new nuclear generation.* Available at www.climatechangecapital.com.

5 Comparing Environmental and Technology Policies for Climate Mitigation and Renewable Energy

CAROLYN FISCHER AND RICHARD G. NEWELL

The potential for renewable energy to displace fossil-fuelled sources of electricity generation has received considerable attention as a means to reduce emissions of greenhouse gases and other pollutants. In 2001, renewable energy sources (for example, geothermal, solar, wind, tide, and hydropower) provided 5.7% of the total primary energy supply for Organisation for Economic Co-operation and Development (OECD) countries (International Energy Agency 2002). For electricity generation, renewables represented 15% of production worldwide, but only 2.1% if one excludes hydropower. Proposals in the United States aim to increase renewable electricity production to 15% by 2020, and the European Union has a target to produce 22% of electricity and 12% of gross national energy consumption from renewable energy sources by 2010 (IEA 2003). Such ambitious targets emphasize the importance of innovation in lowering the cost of these non-emitting energy sources.

Toward these ends, OECD countries have implemented a wide range of policies to reduce greenhouse gas emissions and stimulate innovation in cleaner technologies. Many of the most important policies employ market-based mechanisms. Policies implemented in OECD countries include the following.

- *A carbon dioxide (CO_2) emissions price* – via either an emissions tax or a tradable emissions permit system – provides incentives to reduce CO_2 intensity (that is, the ratio of CO_2 emissions to economic output) and makes fossil-fuelled sources more expensive than re-

This paper is an abbreviated version of Fischer and Newell (2008).

newables. Several Scandinavian countries – and, most recently, the Canadian province of British Columbia – have implemented CO_2 taxes, and in 2005 the European Union launched a program of tradable CO_2 emissions permits.

- *A tax on fossil-fuelled energy* raises the price received by renewables through higher consumer prices for energy, favouring renewables over fossil-fuelled sources. The United Kingdom, Germany, Sweden, and the Netherlands tax fossil-fuelled sources, in most cases by exempting renewable sources from an energy tax.
- *A tradable emissions performance standard*, or generation performance standard, mandates that the average emissions intensity per unit of output (for fossil-fuelled and renewables generation combined) not exceed a standard. Such policies are considered for energy-intensive industries, such as certain sectors in the United Kingdom's Climate Change Levy.
- *Renewable energy portfolio standards* – also called market share requirements or green certificates – may require either producers or users to derive a certain percentage of their energy or electricity from renewable sources. Such programs have been planned or established in Italy, Denmark, Belgium, Australia, Austria, Sweden, and the United Kingdom, as well as in several states and provinces in the United States and Canada.
- *A production subsidy for renewable energy* boosts the price received by renewables and lowers their effective marginal cost relative to other sources, improving the competitiveness of these sources vis-à-vis fossil fuels. The United States has the Renewable Energy Production Incentive of US$0.019 per kWh, and twenty-four individual states have their own subsidies. Canada has a market incentive program, and several European countries and Korea have production subsidies.
- *Subsidies for R&D investment in renewable energy*, including government-sponsored research programs, grants, and tax incentives, are used to encourage near-term and long-term innovations through targeted research. Major programs exist in the United States, United Kingdom, Denmark, Ireland, Germany, Japan, and the Netherlands.

Although economists typically argue that a direct price for CO_2 (via a tax or tradable emissions permit system) would provide the most efficient incentives for development and use of cleaner technologies, the diversity of the present policies suggests that other forces are at play.

First, emissions pricing policies that risk significantly reducing economic activity among energy-intensive sectors have little political appeal. Second, raising the price of CO_2 can have important distributional consequences, both for owners of fossil-fuelled generation sources and for consumers. Third, innovation market failures, such as knowledge spillover effects, imply that emissions pricing alone will not provide sufficient incentive to improve technologies. In other words, a firm may be unwilling to invest in R&D if competing firms are likely to imitate – and benefit from – its advances; such knowledge spillovers can reduce the ability of the innovating firm to obtain appropriate returns from its R&D investment. Credibility problems may also arise in using a promise of high future emissions prices to boost current innovation because such high prices may no longer be desired if and when the resulting cost reductions for non-emitting energy sources arrive (see Kennedy and Laplantem 1999; Newell, Jaffe, and Stavins 1999). Finally, the innovation process may occur not only through R&D investments but also through the ability of firms to learn from the production and use of new technologies (that is, 'learning by doing'); thus, encouraging renewables output may spur innovation. Consequently, output support and other subsidies – alongside emissions regulations and R&D policies – are often attractive to decision makers.

The environmental economics literature on induced innovation has focused on the role of environmental policy in stimulating innovation in environmentally friendly technologies (Jaffe, Newell, and Stavins 2003), including the effects of energy price changes and regulations on innovations in energy technologies (Newell, Jaffe, and Stavins 1999; Popp 2002) and the efficacy of market-based environmental policies relative to prescriptive regulation in inducing efficient innovation (Downing and White 1986; Magat 1978; Zerbe 1970). Among different market-based instruments, innovation and adoption incentives may differ (for example, Biglaiser and Horowitz 1995; Fischer, Parry, and Pizer 2003; Jung, Krutilla, and Boyd 1996; Milliman and Prince 1989; Requate and Unold 2001). However, these studies have primarily focused on comparing emissions pricing policies, such as emissions taxes and auctioned or grandfathered emissions permits (Goulder and Mathai 2000), rather than evaluating a broader, more pragmatic set of policies, such as those using performance standards and supporting renewable energy.

The treatment of technological change is also important in models assessing the impact of climate change policies (Carraro, Gerlagh, and van der Zwann 2003). Most modelling efforts assume exogenous technological change, which is unaffected by policy. The small number of climate-energy models that incorporate induced technological change (which occurs in response to policy) varies in terms of their technological detail, whether or not they focus on technological progress as a function of learning by doing or R&D investments, and the degree to which they have an economic structure capable of computing effects on economic well-being ('welfare effects') (Popp 2004).

In this chapter we review our research that bridges several areas in the analysis of climate policies and induced innovation. This research considers a broad set of environmental policies, including indirect ones, and incorporates 'knowledge investment' through both learning by doing and R&D. We focus on the six main policy options (identified above) for reducing greenhouse gas emissions in the electricity sector and evaluate their relative performance according to different potential goals: emissions reduction, renewable energy production, R&D, and economic surplus. We also assess how the nature of technological progress – the degree of knowledge spillovers and the degree of innovation occurring through learning by doing or R&D – affects the desirability of different policies.

Model

We developed a unified framework to assess the six aforementioned policy options for reducing greenhouse gas emissions and promoting the development and diffusion of renewable energy in the electricity sector. This focus reflects both the policy attention devoted to this sector, which accounts for two-fifths of CO_2 emissions, and the availability of data to parameterize the model. However, the qualitative results will offer insight for other sectors and for a broad-based program to reduce CO_2 emissions across sectors. We provide a more detailed description of the model in Fischer and Newell (2007 and 2008).

The stylized model (figure 5.1) is deliberately kept simple to highlight key features. It includes two subsectors, one emitting (fossil fuels) and one non-emitting (renewables). We assume that both subsectors are perfectly competitive, such that neither consumers nor electricity generators can influence the price of power, and that they supply an identical product, electricity.[1] Fossil-fuelled production includes a CO_2 intensive technology (coal) and a lower-emitting technology (gas turbines). Coal plants operate primarily as baseload generation; their operating costs

Figure 5.1. Schematic of Model

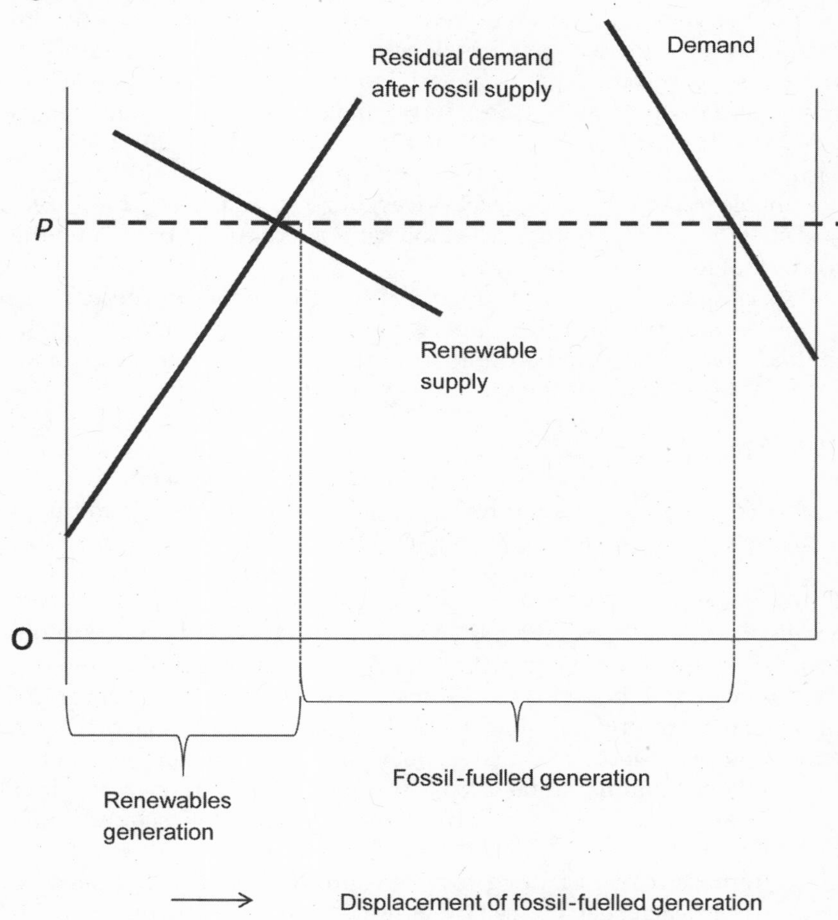

are low, but adding capacity is expensive. In contrast, gas-powered turbines dominate at the margin (that is, where small changes in generation are needed) because they can provide additional capacity at lower cost in response to fluctuating demand. To the extent that renewable energy is competitive, it displaces marginal fossil-fuelled generation. We therefore treat nuclear- and hydro-based generation as fixed in response to the range of policies we model, a reasonable assumption based on other detailed models.

The model has two stages, each representing a specific number of years, to allow for the time required for innovation to occur and for the lifetime of the new technologies. Electricity generation, consumption, and emissions occur in both stages, whereas investment in knowledge takes place in the first stage and, through technological change, lowers the cost of renewables generation in the second. More important, we assume that firms not only take current prices as given (in that they are unable to manipulate market prices) but also take prices in the second stage as given, having perfect foresight about future prices and the forces that influence those prices.

By assuming that no discounting of future benefits to present-day values occurs within the first stage, we ensure constant behaviour within that stage. However, discounting can be allowed within the second, longer stage by altering the duration of that stage accordingly.

The Emitting Fossil-Fuelled Subsector

Each fuel used in the emitting subsector of the generation industry – coal and natural gas – has a fixed CO_2 intensity. The marginal cost (supply) curves for natural gas- and coal-fired generation are assumed to be upward sloping. In other words, the most cost-effective plants are deployed first; as demand increases, production becomes more expensive. The abatement of CO_2 in this subsector relies largely on fuel switching; although coal gasification or generation efficiency improvements are options, they tend to explain little of the predicted emissions reductions in climate policy models (for example, Energy Information Administration 2006). The modest policies we consider would not be sufficient to justify carbon capture and sequestration technologies.

The representative emitting firm maximizes profits with respect to output from each fuel source. We assume that no fuel source is completely driven out of the market in our scenarios and that coal generation is used until its marginal costs (that is, the additional cost for a unit increase in output) are equal to those of natural gas, inclusive of their respective emissions costs. Thus, in the absence of an emissions price, an increase in renewables supply crowds out natural gas- and coal-fired generation in proportion to the slopes of their competing supply curves. A higher emissions price leads to a larger reduction in coal-fired production because this is the only policy that differentiates between coal and gas.

The Non-emitting Renewable Energy Subsector

Another subsector of the industry generates without emissions by using renewable resources. The supply curve for renewable energy sources is assumed to be upward sloping, and increases in accumulated knowledge (or 'knowledge stock') are assumed to lower marginal costs but with diminishing returns. We have simplified considerably by assuming that technological change will occur in the relatively immature renewable energy technologies but not in the relatively mature fossil-fuelled technologies. The incorporation of positive but relatively slower innovation in fossil fuels would complicate the analysis without adding much insight.[2]

Knowledge stock is a function of cumulative knowledge from R&D and cumulative experience through learning by doing. Cumulative R&D-based knowledge increases in proportion to annual R&D-derived knowledge generated in each stage, and cumulative experience increases with total output during the first stage. Research expenditures increase with the amount of new R&D knowledge generated in any one year.

We consider the degree to which innovating firms are able to appropriate returns from their investments in R&D (that is, the degree of appropriability or appropriation rate).[3] In Fischer and Newell (2007 and 2008), we formally derive the appropriation rate for multiple innovating firms, each of which derives its own benefits from knowledge and can, to some degree, appropriate the benefits that accrue to others. From the perspective of our representative firm, appropriated benefits can be represented as a share of the total benefits from R&D. We assume that all knowledge is ultimately adopted, either by imitation or by licensing; therefore, knowledge spillovers do not directly affect the aggregate profit of firms in this subsector. Licensing revenues also do not affect the aggregate profit, because they represent transfers among firms. However, the spillover factor does affect R&D and learning because it determines the share of future profit changes that can be appropriated by the representative innovator.

We take knowledge spillovers into account to determine how the representative firm maximizes profits with respect to output in each stage and R&D investment. The renewable energy subsector produces until the marginal cost of production equals the value it receives from additional output, including the market price, any production subsidy, and the appropriable contribution of such output to future cost reduction through learning by doing. Second-stage output does not generate a learning benefit. Meanwhile, the firm also invests in research until the costs of additional R&D investment equal the expected future returns from that investment.

If appropriation rates are imperfect from a societal perspective (that is, at least some knowledge spillover occurs), firms have insufficient incentive to engage in extra production for the purpose of learning. Similarly, if the R&D subsidy does not fully account for spillover, firms have insufficient incentive to invest in R&D. By discouraging learning and R&D by individual firms, knowledge spillovers reduce overall knowledge in the subsector. Thus, a knowledge externality in the form of knowledge spillovers accompanies the emissions externality (that is, the uncompensated cost to society of CO_2 emissions), and both can be affected by policies that target one or the other.

Consumer Demand

We abstract from short-run peak pricing variations and take a longer-term view of demand and supply curves. Consumer demand for electricity is a function of the price. We model the change in consumer surplus – the difference between the maximum that consumers would be willing to pay and what they actually pay – that is attributable to the policy. In equilibrium, total consumption must equal total supply from fossil-fuelled and renewable energy generation.

Economic Surplus

Policies also have implications for government revenues. The change in revenue transfers equals the tax revenues after accounting for the cost of the subsidies. We assume that the emissions price results in government revenue (through a tax or auctioned permits), whereas a permit system with free allocation would yield the same overall results in our model, with profits accruing to the permit holders rather than government. However, if we assumed that tax revenue increases were used to lower more distortional taxes in the economy, the revenue-raising policy options would perform better, and gaps among policies in efficiency and cost would widen.

Environmental benefits (that is, reduced damages from climate change) are a function of the annual emissions and the length of each stage. The change in economic surplus attributable to policy is the sum of changes in environmental benefits, changes in consumer and producer surplus, and revenue transfers from subsidies or taxes. This includes revenue changes for nuclear and hydropower baseload generators, which have fixed costs and capacities and which profit to the extent that the electricity price rises. In addition to economic surplus, other metrics for evaluating policy may include total emissions, consumer surplus, and renewable energy market share.

To the extent that unmodelled issues – such as leakage of emissions to unregulated generators (Bernard, Fischer, and Fox 2007) or political economy constraints – affect policy performance and welfare impacts, our presentation of economic surplus will not reflect the full social impacts; nonetheless, it is a useful metric.

Optimal Policy

In this model, three policies are needed to optimally address three externalities: emissions, R&D spillovers, and learning spillovers. An appropriate emissions price internalizes the first externality in the sense that emitting firms and consumers bear some of the cost to society of the emissions. To find the optimal subsidies for R&D and learning, we determine how to align the optimal subsidies from the perspective of the firm with those from the perspective of the social planner, whose goal is to maximize welfare. In the presence of spillovers an R&D subsidy is needed to offset the unappropriable share of R&D returns, and a renewables generation subsidy is needed to offset the unappropriated gains from learning.

Policy Scenarios

As we showed in the modelling section, renewable energy production depends on the price received by the renewables subsector (including any subsidies), the cost of R&D investment, and the degree of appropriability of knowledge investments. Fossil-fuelled energy production and the accompanying emissions depend on the level of output from the renewables subsector, the after-tax price of electricity, and the price of emissions. Consumer demand depends on the price of electricity. Different policies affect these prices differentially, resulting in different market equilibria. In this section we use our model to evaluate each policy in terms of the cost to achieve emissions reductions and the extent to which it promotes energy conservation and renewables generation.

No Policy

In both the fossil-fuelled and renewable energy markets and in each stage, output prices equal the baseline price of electricity generation in the absence of policy. We assume that some renewable energy is viable without any policy. Marginal production costs for renewable energy that are less than the price of electricity generation in the first stage would allow for renewable energy production. However, even if marginal production costs are higher than the price in the first stage, renewable energy production could occur if learning by doing lowers second-stage costs. Therefore, even without policy we would expect baseline prices to decline over time as a result of innovation.

Fixed-Price Policies

We first consider three policies that directly set prices: an emissions price, a renewable energy production subsidy, and a tax on fossil-based production. All three policies increase prices for renewables, which expands production and induces more innovation. They differ in their effects on fossil energy production and emissions reduction.

Emissions price. A direct price for emissions provides an incentive for the fossil-fuelled subsector to switch away from coal-fired generation. The market price of electricity increases with the average emissions charge on fossil-fuelled generation. Without other subsidies, the renewables subsector receives the market price for electricity, and the price increase promotes greater renewable energy generation in both stages. The prospect of more output in the second stage increases knowledge-investment incentives in the renewables subsector for both R&D and learning. The higher market price also gives consumers added incentive to conserve. The price increase does not fully compensate for the increase in the marginal costs of coal-fired generation (because lower-emitting gas-fired generation is less costly than coal-fired generation), resulting in less use of coal. In contrast, gas-fired generation may expand if that option is more cost-effective than additional conservation or increased renewable energy generation. In the absence of knowledge spillovers, the emissions price provides efficient incentives for achieving a given emissions reduction goal because it equalizes the incentives for emissions reduction along all three dimensions – reduced emissions intensity, demand conservation (via an electricity price increase), and increased renewables output. With spillovers, however, a price reflecting the additional damage from continued emissions would not offer efficient incentives for knowledge formation.

Renewable energy production subsidy. By putting downward pressure on the electricity price, a renewables production subsidy crowds out fossil-fuelled generation in both stages to reduce emissions. Because the market price of electricity declines, electricity use actually increases. Also, without a direct price on emissions, this policy does not discriminate between coal-fired and gas-fired generation to reduce emissions intensity.

Fossil-fuelled power output tax. A fossil-fuelled production tax raises the price received by renewables by increasing consumer prices for electricity. Thus, both the market price and the effective price received by renewables rise in proportion to the tax. No incentive reduces output or emissions specifically from coal relative to gas plants; however, to the extent that higher prices reduce demand, fossil-fuelled output and emissions will be lower than under an equivalent renewable energy production subsidy.

R&D Subsidy for Renewable Energy Technology

Without a price on emissions or a tax or subsidy on output, an R&D subsidy for renewable energy technology affects output prices in both markets only indirectly. The primary effect of this policy is to increase research expenditures and lower future renewables costs, crowding out fossil-fuelled generation in the second stage. An R&D subsidy provides no incentive for energy conservation or for a reduction in coal generation relative to natural gas generation.

Rate-Based Policies

Portfolio standards and tradable performance standards are two rate-based policies familiar to the electricity generation sector. In effect, both policies create taxes on fossil-fuelled generation and subsidies for renewable energy sources. However, those prices are not fixed but, rather, adjust endogenously, according to market conditions, to achieve the targeted rate.

Endogenous prices raise additional issues with respect to innovation incentives. Essentially, as increased knowledge lowers the costs of renewables, a given standard becomes less costly to meet; this is then reflected in the implicit taxes and subsidies. Innovation incentives differ depending on the structure of markets for output and for innovation (Biglaiser and Horowitz 1995; Fischer, Parry, and Pizer 2003; Jung, Krutilla, and Boyd 1996; Kennedy and Laplantem 1999; Milliman and Prince 1989; Petrakis and Xepapadeas 1999; Requate and Unold 2001).

The question is how the renewables subsector perceives such price changes. We assume that renewable energy firms have perfect foresight about price changes and take them as given. That is, each firm expects knowledge to accumulate and permits prices to respond to that accumulation, but no individual firm expects to influence future prices through its own R&D or learning. In recognizing the equilibration of future prices, R&D incentives are diminished relative to a situation in which firms' expectations are myopic. If firms expect their own actions to further reduce those prices, the expected returns to knowledge will also fall. This contrasts with Fischer, Parry, and Pizer (2003), where the innovators are permit buyers, not sellers. In our numerical simulations, we assume that the rate-based targets are adjusted so that the equilibrium price premium to renewables remains constant over time.

Renewable energy portfolio standard. We model the portfolio constraint in each stage and assume that responsibility lies with the emitting industry to satisfy this constraint. Thus, the fossil-fuelled energy producer must purchase or otherwise ensure a share of at least α_t units of renewable energy for every $(1 - \alpha_t)$ units of fossil-fuelled generation, or $\alpha_t/(1 - \alpha_t)$ green certificates for every unit generated. We allow the standard to tighten over time to facilitate comparison with the price mechanisms and to reflect actual policy proposals better. In equilibrium, the incentives are equivalent to a combination of a subsidy for renewables (the credit price s_t) and a tax per unit of fossil-fuelled output $(s_t\alpha_t/(1 - \alpha_t))$. The implicit tax and subsidy and the electricity price are determined competitively by the market to meet the portfolio constraint.

The portfolio standard provides no incentive to reduce coal reliance relative to gas but crowds out fossil-fuelled generation by implicitly taxing it and subsidizing renewables. Because it combines a fossil energy output tax (which raises electricity prices) with a renewables production subsidy (which lowers electricity prices), the portfolio standard has an ambiguous effect on consumer prices, resulting in limited energy conservation incentives, if any (Fischer 2006). Moreover, for a given portfolio standard, the implicit tax and subsidy decline with reductions in renewable energy costs. This occurs because the implicit tax and subsidy reflect the effective cost of meeting the renewables production constraint, and this 'shadow cost' declines as the cost of renewables production declines. Therefore, to maintain constant policy prices over time, the portfolio standard must become more stringent.

Emissions performance standard. With a tradable performance standard of $\bar{\mu}_t$, the emitting firm must buy emissions permits to the extent that its emissions rate exceeds the mandated standard. The price of emissions, τ_t, will be determined by a market equilibrium. All firms are, in effect, allocated $\bar{\mu}_t$ permits per unit of output, which leads to an implicit subsidy for renewables generation and for lower-emitting fossil-fuelled generation of $\tau_t \bar{\mu}_t$ per unit of output, as they are allocated permits to sell into the market. The fossil-fuelled subsector will be a net buyer of permits costing $\tau_t(\mu_x - \bar{\mu}_t)$ per unit of output for coal generation, x, and $\tau_t(\mu_y - \bar{\mu}_t)$ for natural gas generation, y. (However, if the emissions intensity of natural gas is less than the standard, $\mu_y - \bar{\mu}_t$, gas generation becomes a net seller of permits.) Thus, the emissions performance standard is equivalent to a combination of an emissions price and a generation subsidy for both renewable and fossil energy producers, but the equilibrium values are determined by the market meeting the standard, rather than set directly by policymakers.

The impact of the emissions performance standard on electricity prices could be positive or negative, depending on whether or not the effect of the implicit tax on emissions dominates the effect of the implicit subsidy. The relative effects depend largely on the slopes of the supply curves. Owing to the implicit output subsidy, conservation incentives (if any) are limited; therefore, emissions are higher compared to an equivalent pure emissions price.

Like the portfolio standard, a fixed performance standard implies a subsidy that changes as renewable energy costs fall. An expansion of renewables allows fossil generation emissions to increase – both from greater production and increased emissions intensity as the permit price falls. To maintain constant prices for renewables, the standard must therefore be tightened.

Summary Comparison of Policies

With regard to cost-effective emissions reduction, the relative performance of the policies differs according to the incentives they provide for reducing emissions intensity, reducing energy consumption, increasing renewable energy production, and decreasing costs through innovation (table 5.1; also see Goulder, et al. 1999). Therefore, as one moves from the left of table 1 to the right, efficiency tends to decrease, although the precise ranking of instruments depends on the relative strength of each incentive in particular empirical circumstances.

The revenue and distributional implications of the policies are also quite different. The fossil-fuelled output tax raises revenue, as does the emissions price if it is implemented through auctioned permits or emissions taxes. Price increases are borne by producers and consumers in relation to supply and demand elasticities (that is, the sensitivity of supply and demand to changes in price). The renewables production subsidy and the R&D subsidy require outlays of public funds; taxpayers support the renewable energy producers, whereas electricity consumers and fossil-fuelled energy producers are held harmless. The renewables portfolio and tradable emissions performance standards involve no net revenue change, implicitly earmarking the net costs of these policies back to consumers and producers. The net effects on economic surplus depend on the magnitude of the efficiency loss in the process.

As no one policy perfectly addresses all market failures – the emissions externality and knowledge spillovers from both R&D and learning – a clear ranking cannot be derived analytically. Each policy applies a different set of levers, the relative effectiveness of which depends on parameter values. For example, a tradable performance standard can reduce the emissions intensity of fossil-fuelled generation and subsidize renewable energy, but it discourages conservation. Therefore, it may or may not outperform a fossil output tax or renewables production subsidy alone. Thus, we turn to a numerical application to explore both the magnitude of the efficiency and cost differences and their sensitivity to specific parameter assumptions.

Numerical Application to U.S. Electricity Production

In this section we apply the theory developed above to a stylized representation of the U.S. electricity production sector. We describe the

Table 5.1. Incentives from Alternative Policies

	Emissions price	Tradable emissions performance standard	Output tax on fossil-fuelled generation	Renewables portfolio standard	Renewables production subsidy	Renewables research subsidy
Reduce emissions intensity of fossil fuels	Yes	Yes	No	No	No	No
Energy conservation (via electricity price increase)	Yes	It depends	Yes	It depends	No	No
Direct subsidy for renewable energy output	No	Yes (implicit)	No	Yes (implicit)	Yes	No
Direct subsidy for R&D	No	No	No	No	No	Yes

Table 5.2. Parameter Values and Other Assumptions

Parameter	Base value (US$)
Baseline price of electricity	$0.073/kWh
Intercept of coal-based electricity supply (c_{x1})	$0.023/kWh
Slope of coal-based electricity supply (c_{x2})	$2.2 x 10^{-14} /kWh^2
Intercept of natural gas electricity supply (c_{y1})	$0.061/kWh
Slope of natural gas electricity supply (c_{y2})	$1.8 x 10^{-14} /kWh^2
Intercept of renewable electricity supply (g_1)	$0.059/kWh
Slope of renewable electricity supply (g_2)	$1.2 x 10^{-13} /kWh^2
CO_2 intensity of coal-based electricity (μ_x)	0.96 kg CO_2/kWh
CO_2 intensity of natural gas-based electricity (μ_y)	0.42 kg CO_2/kWh
Electricity demand elasticity (ε)	–0.20
Learning parameter (k_1)	0.15
R&D parameter (k_2)	0.15
R&D investment cost parameter (γ_0)	3.9 x 10^9
R&D investment cost parameter (γ_1)	1.2
Degree of appropriability (ρ)	0.50

results of our central scenario and of sensitivity analyses that further explore key relationships among variables. In Fischer and Newell (2007 and 2008) we specify the types of relationships among predictor and outcome variables in the model. These relationships have the general properties given above, correspond to available information, and are empirically tractable. We also describe the empirical derivation of values for necessary parameters and baseline levels of variables (that is, the levels in the absence of policy) using available information; table 2 in the appendix summarizes the parameter values used in the numerical application.

Policy Simulation Approach

The baseline outcomes of this model in the absence of policy are given in table 5.3. We begin the policy simulations by computing the effects of an emissions price of US$7 per ton of CO_2 (or about US$25 per ton of carbon) throughout the model horizon. For the alternative policies, we compute the level of the policy necessary to achieve the same level of emissions as the emissions price. By holding environmental benefits constant, we can consider differences in economic surplus across policies to arise solely from differences in consumer surplus, producer surplus, and transfers related to electricity consumption and production.

Table 5.3. Baseline Results without Policy

	First stage	Second stage
Price of electricity (US$/kWh)	0.073	0.072
Electricity demand (kWh/year)	4.20×10^{12}	4.20×10^{12}
Coal generation (kWh/year)	2.29×10^{12}	2.27×10^{12}
Natural gas generation (kWh/year)	0.67×10^{12}	0.64×10^{12}
Renewables generation (kWh/year)	0.14×10^{12}	0.20×10^{12}
Nuclear and hydro generation (kWh/year)	1.10×10^{12}	1.10×10^{12}
Renewables share of generation	3.3%	4.8%
CO_2 emissions (billion metric tons CO_2/year)	2.48	2.44
Rate of renewables cost reduction	12%	—

For the portfolio standard and the emissions performance standard, we hold the price of credits constant across stages, while meeting the emissions target. This increases the stringency of these rate-based policies over time (table 5.4). The resulting renewables portfolio standard rises from 6.0% in the first stage to 9.6% in the second stage, which is close to a recent proposal for a national renewables portfolio standard rising from 5% by 2012 to 10% by 2020 (Energy Information Administration 2003).

Central Scenario

The results of the central scenario (see tables 5.4 and 5.5 and figures 5.2 and 5.3) indicate that the emissions price is indeed the most efficient means of achieving a given emissions target, leading to the least cost in terms of surplus and requiring the least investment in renewable energy R&D. An emissions price of US$7 per ton of CO_2 reduces electricity emissions by 4.8% in our central scenario. In the first stage, the CO_2 price reduces coal emissions by 5.7%, encourages switching to natural gas generation (which increases by 7.6%), reduces electricity consumption by 1.0% as a result of a 5.3% rise in the price of electricity, and increases renewables generation by 26%; if one considers both stages, this policy results in the lowest renewables increase of any of the policies. Correspondingly, less renewables R&D investment is necessary than under the other policies. As indicated by the change in consumer surplus, however, the burden on consumers may be large if they do not benefit from revenue transfers.

Owing to the implicit output subsidy for lower-emitting and renewa-

Table 5.4. Annual Effects of Alternative Policies Relative to Base Case

	Emissions price (\$/t CO_2)	Tradable emissions performance standard (tCO_2/GWh)	Output tax on fossil-fuelled generation (¢/kWh)	Renewables portfolio standard	Renewables production subsidy (¢/kWh)	Renewables research subsidy
Policy for 4.8% abatement	7.0	765/743	0.83	6.0%/9.6%	1.4	88%
Electricity price						
1st stage	5.3%	–1.9%	9.4%	–0.4%	–1.5%	–0.2%
2nd stage	5.0%	–2.2%	8.8%	–0.9%	–2.7%	–3.0%
CO_2 emissions						
1st stage	–4.2%	–3.9%	–3.7%	–2.9%	–2.9%	–0.4%
2nd stage	–5.0%	–5.1%	–5.1%	–5.3%	–5.3%	–5.9%
Renewables generation						
1st stage	26%	40%	46%	80%	86%	11%
2nd stage	33%	50%	58%	101%	109%	122%
Coal generation & emissions						
1st stage	–5.7%	–6.4%	–2.7%	–2.1%	–2.1%	–0.3%
2nd stage	–6.3%	–7.3%	–3.7%	–3.9%	–3.9%	–4.3%
Gas generation & emissions						
1st stage	7.6%	15.8%	–11.5%	–9.0%	–8.9%	–1.1%
2nd stage	5.6%	13.0%	–16.1%	–16.8%	–16.8%	–18.8%
Total electricity generation						
1st stage	–1.0%	0.4%	–1.8%	0.1%	0.3%	0.0%
2nd stage	–1.0%	0.4%	–1.7%	0.2%	0.5%	0.6%

Table 5.4. (*Concluded*)

	Emissions price ($/t CO_2)	Tradable emissions performance standard (tCO_2/GWh)	Output tax on fossil-fuelled generation (¢/kWh)	Renewables portfolio standard	Renewables production subsidy (¢/kWh)	Renewables research subsidy
Renewables R&D increase	75%	118%	139%	255%	277%	4043%
Additional renewables cost reduction	4%	5%	6%	10%	11%	25%
ΔConsumer surplus (US$billions/year)	–15.5	6.2	–27.6	2.1	6.4	5.0
ΔProducer surplus (US$billions/year)	–1.1	–6.5	4.4	–2.6	–2.2	–2.5
ΔTransfers (US$billions/year)	16.4	0.0	22.8	0.0	–4.8	–5.5
ΔSurplus (excluding environmental benefits) (US$billions/year)	–0.24	–0.33	–0.38	–0.48	–0.58	–2.9
ΔSurplus relative to emissions price	1.00	1.41	1.61	2.04	2.47	12.49

Note: All dollars are in U.S. dollars.

Table 5.5. Explicit and Implicit Taxes and Subsidies of Alternative Policies

	Explicit and implicit taxes and subsidies			
Policy	Emissions price ($\$/tCO_2$)	Output tax on fossil-fuelled generation (¢/kWh)	Renewables production subsidy (¢/kWh)	Price received by renewables (¢/kWh)
Fixed-price policies				
Emissions price	7.0	—	—	7.7
Output tax on fossil-fuelled generation	—	0.8	—	8.0
Renewables production subsidy	—	—	1.4	8.6
Rate-based policies				
Emissions performance standard	9.5	–0.7	0.7	7.9
Renewables portfolio standard	—	0.1	1.2	8.5

Note: All dollars are U.S. dollars.

Figure 5.2. Cost of Policy Scenarios Relative to Emissions Price

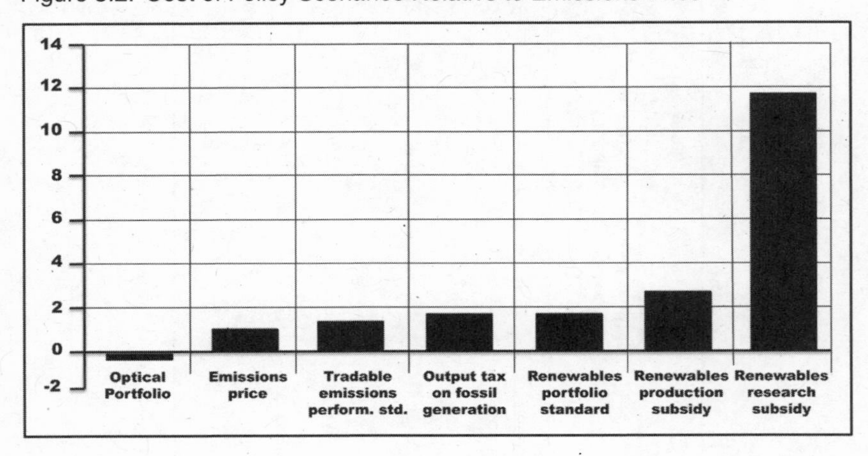

bles generation inherent in the tradable performance standard, conservation plays a lesser role in emissions reductions (consumption actually increases slightly because electricity prices fall). Consequently, more reductions must come from displacing coal and expanding renewables. The implicit emissions price inherent in the tradable performance

Figure 5.3. Increase in Renewable Generation and R&D, Relative to Baseline

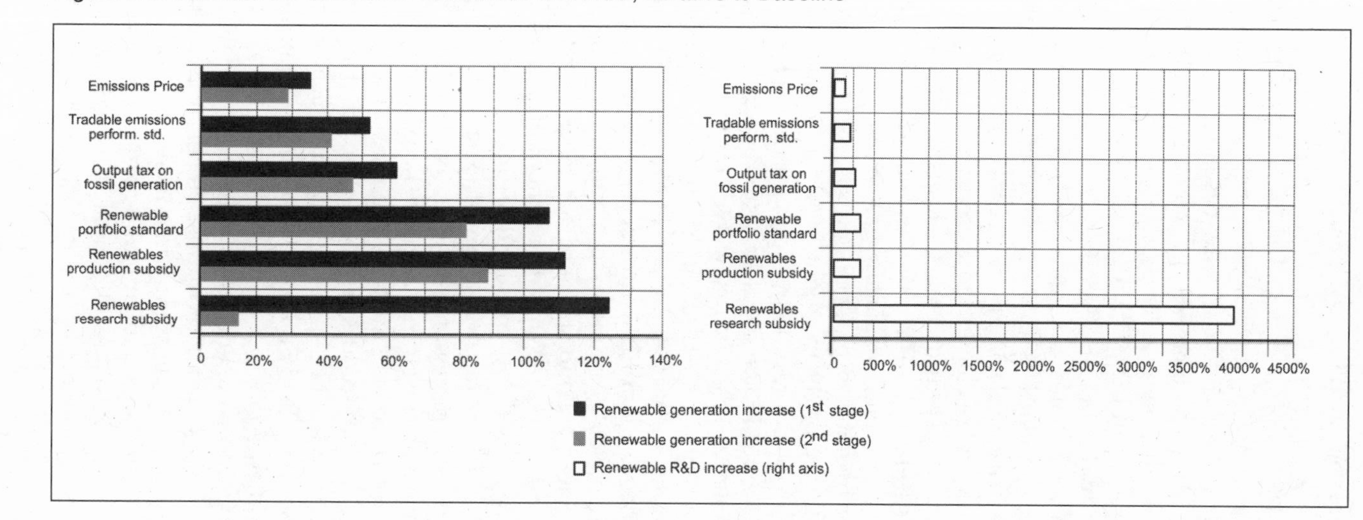

standard is US$9.5 per ton of CO_2, 36% higher than under the explicit emissions price policy. In the first stage, natural gas and renewables replace coal; in the second stage, increased renewables production displaces part of the gas increase. The combination of the implicit emissions price and the renewables production subsidy implies an increase of 0.6 cents/kWh in the price received by renewables, 0.2 cents/kWh higher than under the direct emissions price, inducing moderate expansions in renewables generation over that scenario. The emissions performance standard is the second-most cost-effective instrument. Because the price of electricity falls, consumers directly benefit in this scenario, whereas baseload producers, especially coal, lose revenues.

Because the fossil output tax of 0.8 cents/kWh does not tax emissions directly, this policy places more of the emissions reduction burden on reducing consumption, displacing gas, and increasing renewables; this increases the overall cost of the policy relative to the emissions price. By raising electricity prices by about 75% more than the emissions price, the fossil output tax results in a price increase for renewables of 0.7 cents/kWh, about 75% more than with the emissions price, resulting in comparable increases in renewables output and R&D investment. The fossil output tax almost doubles the costs to consumers, whereas government revenues are significantly higher than under the emissions price. The increased profits to producers using nuclear and hydro power more than offset the profit losses to producers using coal and gas.

A renewables portfolio standard combines an implicit fossil output tax of nearly 0.1 cents/kWh with a renewables production subsidy of 1.2 cents/kWh. With no conservation incentives and no fossil-fuel-switching incentives, renewables must expand three times as much as with the emissions price to meet the target. As a result, its performance is more costly than the emissions price. Because no revenues are raised and electricity prices fall slightly, producers bear the policy burden; renewable electricity producers gain while other producers lose. Consumers also gain from the price reduction, which occurs when the implicit production subsidy outweighs the implicit tax on non-renewable electricity generation.

To result in the same emissions reductions, the renewables production subsidy of 1.4 cents/kWh must be nearly twice the fossil output tax because electricity consumption is increased as a result of a decrease in prices. The effects are similar to the portfolio standard, though somewhat larger, because renewables must also accommodate the increased electricity consumption. Consumers face lower electricity costs, and renewable electricity producers gain, but this is more than offset by the cost to taxpayers and non-renewable electricity producers.

The renewables research subsidy is by far the most costly single policy for reducing emissions. Because cost-effective early emissions reductions are forgone, all emissions reductions must be gained in the second stage by making renewables less expensive than fossil fuels without any emissions reduction or conservation incentives; this requires the cost of renewables to fall 25%. Although second-stage renewables output expands by 122%, it also expands somewhat in the first stage because of the complementarity of learning with R&D.

Figure 5.2 displays the relative costs of the policies, in terms of the change in total surplus, as a ratio to the cost of the emissions price. The optimal combination of policies, which we discuss in a subsequent section, actually leads to a small cost savings with this modest emissions target. The renewable portfolio standard is roughly twice as costly as the emissions price, with the performance standard and the output tax lying in between. The renewables production subsidy is 2.5 times as costly, while relying on the R&D subsidy alone leads to a loss of over 12 times the cost of the emissions price.

Part of the increasing costs of the policies is the increasing reliance on renewables generation to achieve the emissions reductions, as shown in figure 5.3. Research and development also increase with greater reliance on renewables, but relying on the research policy alone requires massive increases in R&D.

The policies also have different implications for the distribution of the costs. Figure 5.4 depicts the change in surplus for consumers, producers, and taxpayers for each policy.

Sensitivity Analyses

The results of several sensitivity analyses are given in table 5.6, where for brevity we focus on the economic surplus of each policy relative to an emissions price of US$7/ton CO_2, as in the bottom row of table 5.4 for our central scenario.

First, we vary the responsiveness of demand, by making the elasticity of demand greater than in the central scenario and then fixing the demand. As illustrated in figure 5.5, a more elastic demand response worsens the performance of all policies except the fossil-fuelled generation output tax relative to an emissions price. This occurs because the reduced electricity prices under the other policies increase demand;

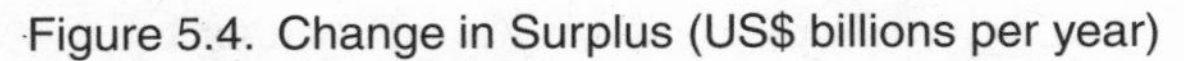
Figure 5.4. Change in Surplus (US$ billions per year)

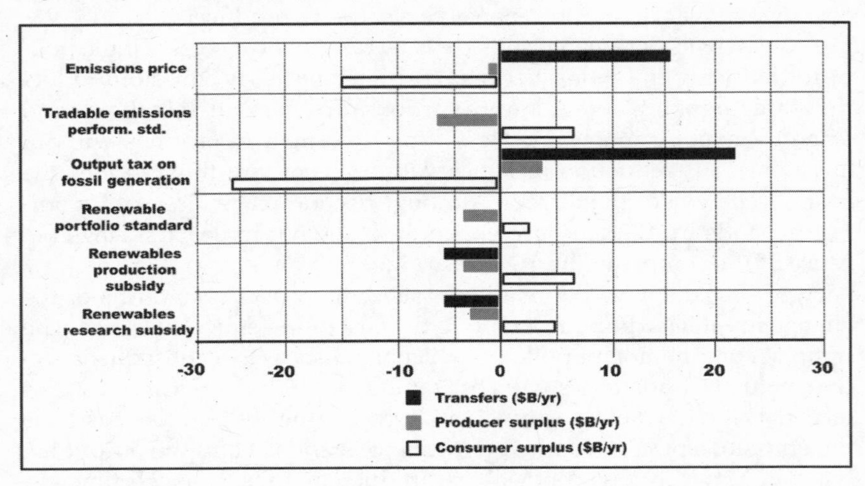

in contrast, the conservation incentives of the fossil-fuelled generation output tax are enhanced. Because conservation does not occur under totally inelastic demand, this scenario illustrates certain equivalencies across different policies with respect to incentives not related to conservation. The emissions price and the performance standard – the two policies that embody incentives for reducing the emissions intensity of fossil fuels – become equivalent because the performance standard no longer relies on a higher implicit emissions price to make up for the extra demand induced by the implicit output subsidy. Similarly, the fossil-fuelled generation output tax, the renewables portfolio standard, and the renewables production subsidy become equivalent because, without conservation, these three policies reduce emissions solely through an increase in the price received by renewables.

We also explore several sensitivity analyses focused on technological change. In the first two, we vary the underlying source of technological change: one in which all technological progress derives from R&D, and another in which learning by doing is relatively more important than in our central scenario. We maintain the same rate of baseline technological change as in our central scenario by adjusting the relative effectiveness of learning versus R&D. The absence of learning by doing makes all but the R&D subsidy perform worse than the emissions price. Without the learning by doing, policies that rely on expanding renewables

Table 5.6. Sensitivity Analyses

Δ Surplus relative to emissions price	Emissions price	Tradable emissions performance standard	Output tax on fossil-fuelled generation	Renewables portfolio standard	Renewables production subsidy	Renewables research subsidy
Central scenario	1.00	1.41	1.61	2.04	2.47	12.49
More elastic demand ($\varepsilon = -0.5$)	1.00	2.05	1.49	2.60	4.12	31.63
Inelastic demand ($\varepsilon = 0.0$)	1.00	1.00	1.55	1.55	1.55	5.45
No learning/higher R&D ($k_1 = 0.0$; $k_2 = 0.19$)	1.00	1.44	1.68	2.26	2.78	4.27
High learning/lower R&D ($k_1 = 0.2$; $k_2 = 0.13$)	1.00	1.40	1.59	1.99	2.41	18.78
High R&D productivity/ baseline learning ($k_1 = 0.15$; $k_2 = 0.20$)	1.00	1.27	1.43	1.27	1.73	4.52
Low R&D productivity/ baseline learning ($k_1 = 0.15$; $k_2 = 0.10$)	1.00	1.50	1.74	2.61	3.09	44.46
Low appropriability ($\rho = 0.25$)	1.00	1.41	1.61	1.93	2.34	3.49
High appropriability ($\rho = 1.0$)	1.00	1.44	1.66	2.44	2.79	18.62
Shorter second stage ($T = 10$, $n_2 = 6.1$)	1.00	1.47	1.70	2.40	2.86	19.07
Longer second stage ($T = 80$, $n_2 = 10.0$)	1.00	1.35	1.53	1.71	2.12	7.95
Flat natural gas electricity supply ($cc_2 = 0$)	1.00	1.50	3.32	4.17	4.83	37.02

Figure 5.5. Sensitivity of Cost Ratios to Changes in Supply and Demand Slopes

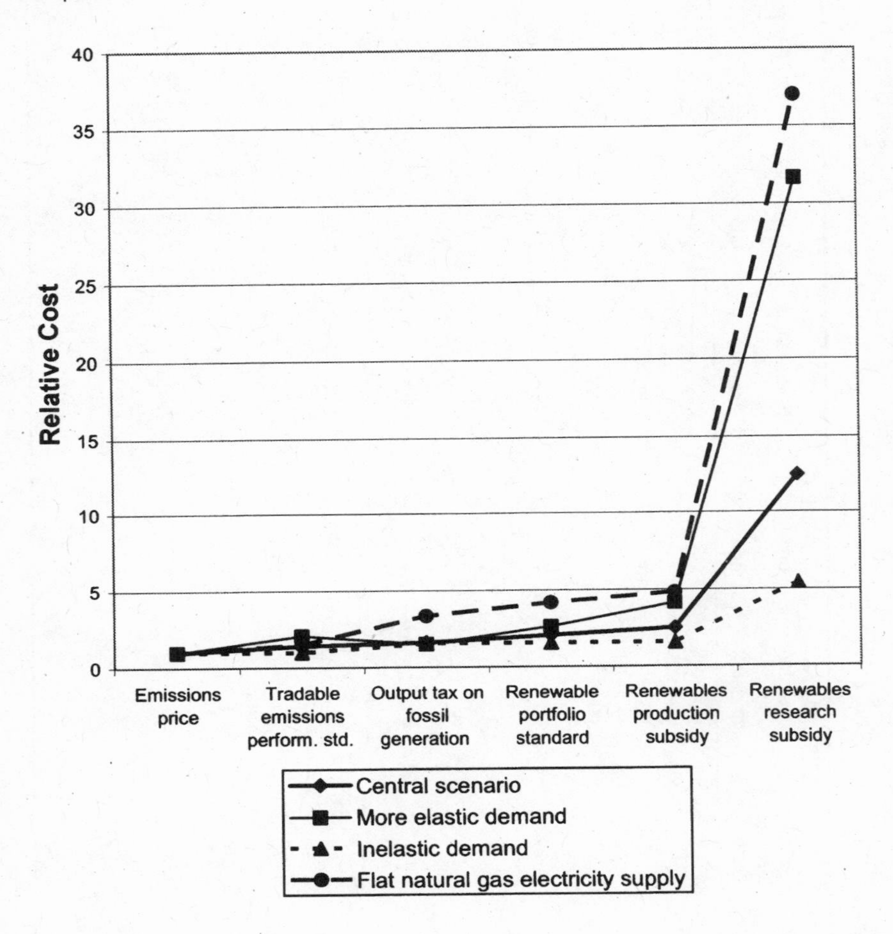

beyond the competitive level of production simply raise costs rather than compensate for spillovers. The 'high learning' case has the opposite effects. In two other closely related scenarios, we alter the productivity of R&D-based knowledge in lowering costs without changing learning. This alters the overall rate of technological change, but the relative performance of policies is affected similarly to the above scenarios that hold the rate of innovation constant.

We evaluate the effect on relative policy costs of the degree of knowledge appropriability in the case of (1) relatively low appropriability and (2) perfect appropriability. Because lower appropriability implies larger knowledge spillovers, the relative performance of any policy that directly or indirectly invests more in knowledge than in the emissions price will improve as spillovers increase. Thus, with low appropriability, the costs of all policies decrease relative to the emissions price, albeit modestly, with the R&D subsidy experiencing the largest improvement because of its direct effect on knowledge investment. The scenario with complete appropriability eliminates the knowledge externality entirely, thereby worsening the relative performance of all other policies, particularly of the R&D subsidy.

We also explore the effect of changing the duration of the second stage. Shortening the second stage from twenty to ten years worsens the performance of all policies relative to the emissions price because it allows less time in the future to benefit from the application of technological change, which is relied on more heavily by policies other than the emissions price. Alternatively, increasing the weight on the second stage (by quadrupling the number of years to eighty) has the opposite effect. In both of these sensitivity cases, however, the policy ranking remains the same.

The final sensitivity analysis examines a flat natural-gas electricity supply function. In this case, additional renewables generation displaces only natural-gas-fired generation, not the more emissions-intensive coal-fired generation, making it more difficult to achieve reductions.

Combination Policy for Addressing Technology Spillovers

The presence of knowledge spillovers in our model implies that separate policy instruments are necessary to optimally correct the climate externality and the externalities associated with both learning by doing and R&D. In this section, we explore a combination of policies to empirically assess the theoretical findings summarized in table 5.1, while maintaining all assumptions at their values in our central case.

We assess a scenario with a portfolio that includes an emissions price, a learning subsidy (that is, renewables generation subsidy), and an R&D subsidy. The learning subsidy and R&D subsidy are set at their optimal levels, and the emissions price is set at a level such that this portfolio of policies achieves the same emissions target as does an emissions price of US$7/ton CO_2 alone (that is, a 4.8% reduction). The optimal R&D subsidy of 50% is straightforward to compute (see Fischer and Newell 2007, 2008) based on the assumed degree of appropriability. The optimal first-stage 'learning subsidy' for renewables depends on the future level of renewables use and other variables; it must therefore be solved numerically. As described in Fischer and Newell (2007, 2008), we find that the optimal learning subsidy is 0.3 cents/kWh, or 4% of the electricity price. Although this subsidy for learning may appear small, it is similar to the optimal R&D subsidy if both are expressed as a percentage of revenue; that is, the optimal R&D subsidy of 50% has a corresponding R&D level that is 12% of renewables revenue, such that the R&D subsidy is 6% of revenue. Nonetheless, this level of learning subsidy is much lower than the levels typically seen in practice, suggesting that, for relatively mature renewables in particular (for example, wind), it is difficult to rationalize large subsidies on the basis of learning.

As a result of greater cost reductions and renewables penetration with the R&D and learning subsidies, the emissions price necessary to achieve a 4.8% emissions reduction falls by 36% to US$4.5/ton CO_2 (from US$7.0/ton CO_2; figure 5.2). Importantly, the associated cost of the combination policy actually becomes negative and instead has a slight annual surplus of US$0.04 billion. We find that the net surplus of the combination policy is attributable primarily to the R&D subsidy (US$0.20 billion surplus annually) offsetting the cost of the emissions price, rather than to the learning subsidy (US$0.02 billion surplus annually). However, the optimal R&D and learning subsidies lead to emissions reductions of only 1.4% and 0.3%, respectively. These results demonstrate that the cost of emissions reductions through a combination of policies could be considerably lower than the cost under policies targeted solely at technology R&D or emissions pricing. Nonetheless, the emissions reductions associated with this policy package are due primarily to the emissions price.

Conclusion

In our comparison of policy options for reducing CO_2 emissions and promoting renewable energy, some clear principles emerge. When the ultimate goal is to reduce emissions, policies that create incentives for fossil-fuelled energy generators to reduce emissions intensity and for consumers to conserve energy perform better than those that rely solely on incentives for renewable energy producers. We find that the nature of knowledge accumulation – whether through learning by doing or R&D – is far less important than the nature of the policy incentives. This is consistent with Parry, Pizer, and Fischer (2003), where the welfare gains from correcting environmental externalities are more important for policy than are the gains from technological change. Nonetheless, the nature of technological change and the degree of knowledge spillovers do have discernible effects on the relative cost of policies that differentially affect knowledge investment and how it occurs.

For the moderate emissions targets we explore, a renewable energy R&D subsidy turns out to be a particularly inefficient means of emissions reduction because it postpones the vast majority of the effort to displace fossil-fuelled generation until after the costs of renewables are reduced (see also Schneider and Goulder 1997). The use of a renewables R&D subsidy alone would require very large R&D investments and forgoing near-term cost-effective abatement opportunities. Although climate change is a long-term problem, our results for mid-term strategies emphasize the important role of policies that encourage abatement across all available forms of energy generation and time frames, as well as the limitations of narrowly targeted policies.

We show that an emissions price alone is the most efficient single policy for reducing emissions because it simultaneously gives incentives for fossil energy producers to reduce emissions intensity, for consumers to conserve, and for renewable energy producers to expand production and to invest in knowledge to reduce their costs. The other policies offer different combinations of these incentives with correspondingly different consequences for the distribution and the overall size of the burden of meeting an emissions reduction target.

Nonetheless, no single policy can simultaneously correct more than one market failure – an emissions externality and knowledge spillovers from learning and R&D. Each policy then poses different tradeoffs (Jaffe, Newell, and Stavins 2005). In the presence of knowledge spillovers, we find that an optimal portfolio of policies – an emissions price combined with optimal learning and R&D subsidies – can achieve emissions reductions at significantly lower cost than can any single policy alone (as in Schneider and Goulder 1997), although the emissions reductions continue to be attributable primarily to the emissions price, and the optimal learning subsidy is small.

Together, these results call into question some of the arguments by Montgomery and Smith (2007), who suggest that R&D is the key for addressing climate change and that an emissions price high enough to induce the needed innovation cannot be credibly implemented. Although a high future emissions price may not be credible, the combination policy allows for a much more modest emissions price. If one believes that even a modest emissions price is not politically feasible, an R&D subsidy by itself is not the next best policy, and the costs of that political constraint are likely to be quite large and increasing with restrictions on the remaining policy options. It should be kept in mind, however, that we focus on reductions over the near- to mid-term and incremental improvement of existing technology, rather than the development of breakthrough technologies that might achieve deep reductions. Research and development policies probably have greater salience in the latter context, although this lies beyond the scope of this chapter.

ACKNOWLEDGMENTS

We acknowledge financial support from the U.S. Environmental Protection Agency and the Swedish Foundation for Strategic Environmental Research (Mistra).

NOTES

1 Although large portions of the electricity sector remain regulated, policy-induced changes to marginal production costs are likely to be passed to consumers; in a longer horizon, a transition to deregulated markets is likely to make markets relatively competitive in the future.
2 An exception is advancement in cleaner fossil generation technologies, like carbon capture and storage. Our qualitative results should carry over to policies targeting other low-carbon technologies, although the quantitative results would depend on the cost, technology, and emissions parameters particular to those other technologies.
3 We model general knowledge as being appropriable by the innovating firm, with no distinction regarding the source of knowledge (R&D or learning by doing). Although an empirical basis is lacking for such a distinction, one might expect some forms of learning to be less easily appropriated by other firms. For example, if learning is more firm-specific and less likely to spill over, policies subsidizing renewables are less appropriate to compensate for knowledge externalities. In contrast, if learning is more difficult to patent in order to appropriate rents, then renewable subsidies may be relatively more justified.

REFERENCES

Bernard, A., C. Fischer, and A. Fox. 2007. 'Is there a rationale for output-based rebating of environmental levies?' *Resource and Energy Economics* 29(2): 83–101.

Biglaiser, G., and J.K. Horowitz. 1995. 'Pollution regulation and incentives for pollution control research.' *Journal of Economics and Management Strategy* 3:663–84.

Carraro, C., R. Gerlagh, and B. van der Zwann. 2003. 'Endogenous technological change in environmental macroeceonomics.' *Resource and Energy Economics* 25(1): 1–10.

Downing, P.G., and L.J. White. 1986. 'Innovation in pollution control.' *Journal Environmental Economics and Management* 13:18–29.

Energy Information Administration (EIA). 2003. *Analysis of a 10-percent renewable portfolio standard*. Washington, DC: EIA.

– 2006. *Energy market impacts of alternative greenhouse gas intensity reduction goals*. Washington, DC: EIA.

Fischer, C. 2006. 'How can renewable portfolio standards lower electricity prices?' Resources for the Future (RFF) Discussion Paper 06-20. Washington, DC: RFF.

Fischer, C., and R.G. Newell. 2007. 'Environmental and technology policies for climate mitigation.' Resources for the Future (RFF) Discussion Paper 04-05 (revised). Washington, DC: RFF.

– 2008. 'Environmental and technology policies for climate mitigation.' *Journal of Environmental Economics and Management* 55(2): 142–62.

Fischer, C., I.W.H. Parry, and W.A. Pizer. 2003. 'Instrument choice for environmental protection when technological innovation is endogenous.' *Journal of Environmental Economics and Management* 45(3): 523–45.

Goulder, L.H., and K. Mathai. 2000. 'Optimal CO_2 abatement in the presence of induced technological change.' *Journal of Environmental Economics and Management* 39:1–38.

Goulder, L.H., I.W.H. Parry, R.C. Williams, and D. Burtraw. 1999. 'The cost-effectiveness of alternative instruments for environmental protection in a second-best setting.' *Journal of Public Economics* 72:329–60.

International Energy Agency (IEA). 2002. 'Renewables in global energy supply.' IEA Fact Sheet (November). Paris: IEA.

– 2003. *Renewables information*. Paris: IEA.

Jaffe, A.B., R.G. Newell, and R.N. Stavins. 2003. 'Technological change and the environment.' In *Handbook of environmental economics*, vol. 1, ed. K.-G. Mäler and J.R. Vincent, 461–516. Amsterdam: Elsevier.

– 2005. 'A tale of two market failures: Technology and environmental policy.' *Ecological Economics* 54:164–74.

Jung, C., K. Krutilla, and R. Boyd. 1996. 'Incentives for advanced pollution abatement technology at the industry level: An evaluation of policy alternatives.' *Journal of Environmental Economics and Management* 30:95–111.

Kennedy, P.W., and B. Laplantem. 1999. 'Environmental policy and time consistency: Emissions taxes and emissions trading.' In *Environmental regulation and market power: Competition, time consistency and international trade*, ed. Petrakis, Sarzetakis, and Xepapadeas. Northampton, MA: Edward Elgar Publishing.

Magat, W.A. 1978, 'Pollution control and technological advance: A dynamic model of the firm.' *Journal of Environmental Economics and Management* 5:1–25.

Milliman, S.R., and R. Prince. 1989. 'Firm incentives to promote technological change in pollution control.' *Journal of Environmental Economics and Management* 17:247–65.

Montgomery, W.D., and A.E. Smith. 2007. 'Price, quantity and technology strategies for climate change policy.' In *Human-induced climate change: An interdisciplinary assessment*. Cambridge: Cambridge University Press.

Newell, R.G., A.B. Jaffe, and R.N. Stavins. 1999. 'The induced innovation hypothesis and energy-saving technological change.' *Quarterly Journal of Economics* 114:941–75.

Parry, I.W.H., W.A. Pizer, and C. Fischer. 2003. 'How large are the welfare gains from technological innovation induced by environmental policies?' *Journal of Regulatory Economics* 23(3): 237–55.

Petrakis, S., and A. Xepapadeas, eds. 1999. *Environmental regulation and market power: Competition, time consistency and international trade*. Northampton, MA: Edward Elgar Publishing.

Popp, D. 2002. 'Induced innovation and energy prices.' *American Economic Review* 92(1): 160–80.

– 2004. 'ENTICE: Endogenous technological change in the DICE model of global warming.' *Journal of Environmental Economics and Management* 48:742–68.

Requate, T., and W. Unold. 2001. 'On the incentives created by policy instruments to adopt advanced abatement technology if firms are asymmetric.' *Journal of Institutional and Theoretical Economics* 157(4): 536–54.

Schneider, S.H., and L.H. Goulder. 1997. 'Achieving low-cost emissions targets.' *Nature* 389:13–14.

Zerbe, R.O. 1970. 'Theoretical efficiency in pollution control.' *Western Economic Journal* 8:364–76.

6 A Discussion of Electricity's Role in Reducing the Environmental Footprint of Energy Use

JAMES MEADOWCROFT

These comments are organized under three headings: the two contributions to the session entitled 'Electricity's role in reducing the environmental footprint of energy use'; Ontario's electricity supply system and climate change; and electricity's role in reducing CO_2 emissions.

The Two Contributions

Carolyn Fischer and Richard Newell's chapter explores options for CO_2 abatement by modelling the impact of six policy designs in the U.S. electricity sector. Perhaps unsurprisingly, they find that the lowest cost option for achieving a specified emissions reduction is an emissions price (which could be provided through either a carbon tax or a cap-and-trade system). The higher costs encourage a range of actors to use multiple pathways to reduce emissions. The most expensive single policy option turns out to be R&D subsidies, as huge amounts of money must be directed at the renewables sector to accelerate deployment of these technologies. In fact, a mix of policies performs better than any individual policy, because multiple instruments are required to handle multiple externalities (those associated with R&D spillovers and learning spillovers as well as with emissions). Thus the addition of (modest) learning and R&D subsidies to an emissions price produces the most cost-effective policy package.

There is much that could be said about the model employed in this discussion, but two words of caution are in order. First, in the real world the obstacles to the penetration of new technologies in the electricity sector are more complex and more significant than those captured in this model. Consider, for example, grid architectures that make it difficult to integrate renewables. Second, the CO_2 reductions modelled here are extremely modest. The central scenario involves a reduction of approximately 5%, consequent on an emissions price of $7 per ton of carbon dioxide. However, scientific evidence suggests that in developed countries reductions of perhaps 20% to 30% by 2025 and 80% or more by 2050 may be required to avoid dangerous climate risks (Intergovernmental Panel on Climate Change 2007). Both considerations suggest that other sorts of policy measures, including those that target the development and deployment of new technologies, will be more critical than the chapter implies (Stern 2006).

Dominique Finon's chapter examines practical experience with climate change policies in the context of liberalizing energy markets within the European Union. He emphasizes the risks that energy producers assume in liberalized markets and the way in which the compounding of additional risks associated with the use of market-based CO_2 abatement instruments (the European Emission Trading System) can raise the costs of capital and generate deleterious consequences. A particularly interesting feature of the chapter is the vigorous defence of feed-in tariffs for developing renewable electricity supply, and the contention that this instrument produces cheaper (as well as more rapid) deployment of renewables than do the rival quotas and tradable green certificates schemes. The chapter illustrates the risks and complexities of emissions trading. Nevertheless, it should be remembered that emissions taxes are also subject to regulatory uncertainty, for taxes passed (or promised) by one government can be undone by a subsequent administration.

Ontario's Electricity Supply System and Climate Change

Three features of Ontario's electricity supply system should be kept in mind when considering climate change policy. First, the existing system results from a stalled attempt to introduce competition and privatize a previously centralized and state-owned sector. Its current configuration is not what anyone intended or anticipated. Despite the free-market trappings, central political control of the system has been reaffirmed – although this control is exercised indirectly through various bodies over which the Ontario government has different degrees of direct influence. From a period of high price volatility we have now entered one of relative stability where supply is mostly governed by long-term contracts. The institutional structure remains fragmented, with various bodies (Independent Electricity System Operator, Ontario Power Authority, Ontario Energy Board, and Hydro One) having different operational mandates. There is more plan than market, but the planning system is not one that a 'rational planner' would design.

Second, until recently, developments were largely driven by a series of crises: the shutdown of much of the nuclear plant in the 1990s; price increases and volatility that upset consumers and worried politicians; then blackouts, brownouts, and supply concerns. The political commitment to phase out coal brought a new round of supply anxieties, and a great deal of effort in the last few years has gone to securing electricity supply for the coming decade.

Third, climate policy has not been a significant concern in the electricity sector. Furthermore, to the extent that the Ontario government began to develop such a policy, it rested largely on measures that had been adopted for other reasons. Thus, the cornerstones of the government's approach were (1) the coal phase-out, which was introduced as a political pledge by the current premier, while he was in opposition, to address concern over criteria (air) pollutants, and (2) energy efficiency and demand management measures, which were adopted to deal with supply constraints and reliability concerns.

Electricity's Role in Reducing CO_2 Emissions

Electricity has the potential to play a critical role in reducing carbon emissions in Ontario, and globally. Electricity generation now accounts for 17% of Ontario's greenhouse gas emissions (Environment Canada 2005), and the continued presence of coal (20.3% of electricity supply) (Independent Electricity System Operator 2009) and natural gas (10.7%) (IESO 2009) in the generating mix provides significant opportunities for emissions reduction. However, for this abatement potential to be realized, climate change must be taken seriously by policymakers, and it must become central to decisions about the future design of the electricity system.

Taking climate change seriously has a number of implications. First, it suggests that scientific understanding of the scale of the climate risk must serve as a critical reference point for decision making. Recent assessments point to the dramatic scale of the necessary reductions – perhaps 80% or more of CO_2 emissions in developed countries by 2050. The high speed and massive scale of these anticipated reductions provide a critical context for decision making in the electricity sector.

Second, climate policy should be explicit, not implicit, and it is an issue with which all actors in the electricity system must engage. As we have seen, the main pillars of Ontario's climate policy in the electricity sector were put in place for other reasons. This is a perfectly understandable way of initiating reduction efforts. It is no accident that climate leaders in Europe include countries that secured emission reductions because of a fortuitous turn of events or as a by-product of policies oriented towards other ends – Germany, with the decline of industry in the east after reunification, and the United Kingdom with the 'dash for gas' (Margaret Thatcher's revenge on the coal miners). In order to move beyond such windfalls, emissions reductions must become an important and explicit policy orientation for government. It seems that the Ontario government is headed in this direction, with the new energy and infrastructure minister emphasizing the importance of climate change in electricity decision making. However, it is also important that key electricity players (regulatory agencies and operating entities) have climate change and emissions abatement clearly stipulated as part of their statutory concerns. This is so important, because without an explicit focus on climate impacts, institutions are likely to make different choices.

Third, short-term emissions reductions matter. There are two reasons this is so. Total emissions – rather than just emissions at some particular point in the future – are significant for the climate; by starting to bend the curve immediately, the overall volume of emissions can be cut more significantly. Secondly, there is the politics of the issue. Short-term reductions demonstrate that the issue is being taken seriously and that politicians are prepared to act today. While promises of future cuts can be more easily discounted as rhetoric, immediate reductions show that developed countries are making good on their commitment to move first, thereby helping to convince developing countries to join a broader international effort to deal with climate change. On this front, Ontario's straightforward plan to phase out coal-fired electricity generation is positive, although some may wonder whether a place for carbon capture and storage might not one day be found in the province's energy mix.

Fourth, decisions about electricity should be set in the context of the broader economy-wide transition to a carbon-neutral economy. To put this another way: there is little point thinking about carbon abatement in the electricity sector without considering reductions across the economy as a whole. This means understanding that the need for electricity may expand dramatically in coming decades as electricity is expected to function more widely as an energy carrier. For example, the Ontario government emphasizes that the Integrated Power System Plan will deliver significant emissions reduction in the period up to 2020 and beyond (Government of Ontario 2007). This sounds good, but the goal should be to *maximize the emissions reduction across the energy system*, not just emissions from the electricity sector. The position of electricity (size and distribution of generating assets, fuel mix, and CO_2 emissions) must be considered in relation to what is going on in other sectors. For example, transport currently accounts for 32% of Ontario greenhouse gas emissions (Environment Canada 2005), and in the short term, abatement alternatives are restricted as the sector is almost entirely dependent on oil. However, low-carbon alternatives under development include biofuels, hydrogen (generated by low-carbon energy sources or fossil fuels employing CCS), all-electric vehicles, and plug-in hybrids. A dramatic shift to hybrids or all-electric vehicles would eliminate emissions from fossil fuels in the transportation system, but it would increase electricity demand – all that power would have to come from somewhere. The point is that planning for the electricity sector (for demand, supply, and emissions abatement) must be integrated with a vision for the overall emergence of a low-carbon energy system by mid-century.

Fifth, decisions about climate technologies should consider the life-cycle analysis of energy production and consumption and not just rely on emissions at the generation facility. The preparation and transport of fuel, for example, can have significant greenhouse gas implications, and this is true for biofuels and nuclear as well as coal. Moreover, climate impacts cannot be abstracted from other sustainability issues such as emission of criteria pollutants, water consumption, and land use.

Several general considerations can be drawn from these two chapters and from the brief remarks above. There is no doubt that carbon pricing is critical to getting things going, by sending a clear signal across the economy that we are entering a carbon-constrained world and that henceforth emitting CO_2 will become increasingly expensive. Of course, every possible system for generating a carbon price (upstream or downstream carbon taxes, upstream or downstream cap-and-trade systems, and so on) carries risks of regulatory failure and/or economic abuse, but that is the way it is with policy design and implementation in any area in the real world; it is no excuse for inaction. The great uncertainties associated with climate change policymaking, and the high (economic and environmental) stakes involved in decisions about the electricity system, suggest that system *flexibility* and *resilience* have enormous value. It is important to keep options open and to be capable of withstanding adverse circumstances and unanticipated shocks. This suggests that more attention should be paid to local options including combined heat and power, biomass facilities, and other emerging renewables. Finally, research, development, and deployment of new carbon-neutral technologies is critical. In this regard, emerging renewables-enabling technologies – such as energy storage, grid architectures, and systems control – have not yet received the attention they deserve.

REFERENCES

Environment Canada. 2005. 'National inventory report, 1990–2005: Greenhouse gas sources and sinks in Canada.' Available at http://www.ec.gc.ca/pdb/ghg/inventory_report/2005_report/tdm-toc_eng.cfm.

Government of Ontario. 2007. 'Ontario greenhouse gas emissions targets: A technical brief.' Available at http://www.ene.gov.on.ca/publications/6793e.pdf.

Independent Electricity System Operator (IESO). 2009. 'Supply overview.' Available at http://www.theimo.com/imoweb/media/md_supply.asp.

Intergovernmental Panel on Climate Change (IPCC). 2007. 'Synthesis report.'

Stern, N. 2006. *The economics of climate change: The Stern review*. London: HM Treasury.

PART THREE

Finding the Right Price

7 Introduction

DONALD N. DEWEES

No discussion of electricity policy is complete without a discussion of the price. Consumers press relentlessly for low, stable prices. Generators insist on prices that cover all costs including a reasonable return on invested capital. Advocates of competitive markets call for prices that reflect supply and demand in order to avoid shortages (brownouts or blackouts) or costly excess capacity. As part 2 shows, price can play an important role in reducing the environmental harm caused by the generation or transmission of power. Part 4 shows that price is an important factor in achieving end-use electricity efficiency goals. Politicians want prices that are competitive with or lower than those in surrounding jurisdictions in order to avoid a competitive disadvantage for domestic industry.

In the past, electricity pricing was relatively simple, and prices were stable. Regulators approved fixed rates that integrated utilities charged to various classes of customers for one or more years. However, technology now allows prices to be set in five-minute intervals, varying by location, by customer, and even by appliance, while competitive markets can reveal market-clearing wholesale spot prices that may reflect short-run marginal costs. The array of pricing options is now vast. In the past, consumers thought little about electricity prices except to complain if they were too high. Now they may be asked to choose among pricing plans with which they have little experience, when they are unable to predict the plan that will be least costly for their consumption pattern. What should policymakers assume about the ability of consumers to understand pricing options, to choose wisely in a competitive pricing market, or to respond wisely to sophisticated pricing regimes?

After decades of experiencing real (inflation-adjusted) electricity prices falling as technology improved, we are now gaining experience with prices rising to reflect higher fuel costs, increased pollution control costs, and 'green' generation costs. Competitive markets have introduced some jurisdictions to volatile and hard-to-predict prices. Planners and market participants want to know where prices are going. Policymakers are confronted with writing the rules that will direct those prices.

The chapters in part 3 help to illuminate the implications of various choices for electricity pricing policy. They also suggest where both policy and prices are likely to go and where they ought to go both in general and in Ontario in particular. Fereidoon Sioshansi argues that Ontario's electricity prices should move higher, not lower, and he explores the damaging consequences that arose in other jurisdictions and other markets when governments did not have the political will to pursue such necessary price increases. He also explores the ways in which we should use the available technology for sophisticated pricing policies. Dean Mountain presents hard evidence from studies in Ontario and elsewhere of the ways in which consumers respond to various price programs.

Existing research helps to identify the price and information programs that may persuade individuals to use electricity more efficiently, matching their consumption to the cost and availability of power. To achieve the maximum impact, prices must be coupled with information about real-time consumption by individual appliances and probably with automatic systems for turning some appliances down or off under certain price and system conditions. Some consumers may want to respond manually to real-time prices. Others may prefer to have their air conditioning or electric hot water or other major appliances adjusted automatically by their supplier under an easy-to-understand protocol. Very few will want to spend much time studying alternative price packages and making optimizing decisions. So, we need price packages that vary according to differing customer load patterns, differing customer desire to be involved in energy management, as well as differing tastes for bearing price risk.

As price packages become more complex, they are harder for customers to evaluate, and customers will more often be misled or disappointed. As price packages become more complex, there are more opportunities for market participants to game the system, imposing externalities on other players. As price packages become more complex, they become more opaque, and it is more likely that influential parties can insert provisions that give them an unfair advantage over small consumers.

How do we find the right set of price/control packages to maximize consumers' control of their consumption while avoiding confusion and its costs? Is it by encouraging marketers to design the packages and encourage consumers to leave their utility in favour of a competitive marketer? Or is it by allowing the regulator to approve a set of packages that the utility offers, leaving competitive retailers to look for niches not supplied by the default supplier? Should the regulator specify the price package that marketers may offer, letting them compete on some price measure, or should the competitive offerings be wide open?

George Vegh reminds us that Ontario has a semi-market in which significant parts of the price are determined by administrative decisions, not by market forces. The government's environmental policy has been to minimize the price impact of environmental improvement, mandating a coal phase-out without raising the price of coal power to reflect its environmental costs. Other sources of CO_2 face no price at all. If we are to be off coal in 2014, we will need increased supply from other sources and a functioning market to stimulate energy conservation and demand management.

All of these challenges arise at a time when the government is reluctant to revisit market design. The last government that radically changed the Ontario electricity system was voted out of office, and it seems unlikely that big changes would be politically attractive to the public in the near future, despite their appeal to experts in the field. Yet some changes in pricing need to be made if Ontario is to come close to reaching its goals. We hope that the search for politically viable pricing policies will benefit from the following chapters in part 3, 'Finding the Right Price.'

8 What Is the 'Right' Price for Electricity in Ontario?

FEREIDOON P. SIOSHANSI

What is the right price for electricity in Ontario? The short answer, economically speaking, is the price that brings supply and demand into balance. In practice, of course, things are never quite that simple. The correct price should be determined in the context of broader objectives and policies, especially in times of dramatic change or discontinuity. This chapter is an attempt to provide more sensible and pragmatic approaches to the question in the context of the future of electricity in the province of Ontario.

Before the discussion begins, however, a few caveats apply and should be spelled out. First, I must point out that I am *not* an expert in the Ontario electricity market and that this chapter does not attempt to get into a debate on the merits or shortcomings of Ontario's current policies and plans.

Second, even without knowing the details of the Ontario case, I venture to say that electricity prices are likely to be going up in Ontario – as in many other places[1] – and consumers and industry should begin to prepare for that eventuality. The broad implications are that consumers have to rationalize the new price regime. Rising energy prices, in my opinion, are not a passing phenomenon and are not limited to electricity or Ontario. The era of cheap and plentiful energy is over as the world faces new constraints including carbon constraints, water shortages, a prolonged commodity boom, and rising construction costs for power generation.

Third, for those who find this message depressing, I propose to make the unpopular argument that for too long Ontario, as in many other places in the world, has had it too good by paying relatively little for electricity, not accounting for its full environmental externalities. Going forward, more of these externalities are likely to be internalized, thereby exacerbating likely price increases.

Fourth, in response to question of what is the right price, this chapter will cite a few examples of wasteful and irrational use of electricity and ways to moderate such use. The so-called prices-to-devices revolution, the introduction of smart metering and smart pricing, promises to deliver additional benefits through automation, bypassing the need to manually respond to changing prices. More sophisticated forms of demand response also promise some relief but are complicated to implement.

Fifth, this chapter offers a number of anecdotes with obvious parallels to what appears to be confronting policymakers in Ontario today. These examples show how other communities have responded to significant price increases, sometimes with happy endings. My suggestion is that these anecdotes are applicable to Ontario, at least in the broadest sense, and that they offer useful insights.

I start by describing the difference between incremental and radical price adjustments and what may be appropriate when there is a major discontinuity. This is followed by a discussion of the importance of price as a conveyor of value. With the passage of time and changes in relative values, sometimes the price of electricity becomes partially or completely disconnected from its value. This may be among the reasons that justify significant price adjustments, even if they are onerous for some consumers. Finally, the chapter argues that sometimes it may be justified to adjust prices to support a broader political agenda. Recent decisions to reduce emissions of greenhouse gases in order to avert global climate change, for example, require significant price increases and other changes. These may not be popular decisions for any politician to take, but they may be necessary given the even more unpleasant alternatives.

Price Influences Demand

We all know that changes in price affect consumption. Let me offer a recent and concrete example of this well-known phenomenon, namely an avalanche cutting off a major power line that feeds the city of Juneau, Alaska (*Wall Street Journal* 2008). The local utility, Alaska Electric Light and Power, announced that it would take three months to repair the line that brings low-cost hydropower to the city. In the interim, the utility had to use expensive diesel generators to meet local demand at an estimated cost of US$25 million. Residents were warned that electricity prices would jump from roughly 10 to 50 cents/kWh. Customers got the message quickly and powered down even before receiving their higher bills. Energy consumption dropped by 30% in mere anticipation of higher prices – plain and simple.

Would the people of Juneau have reacted so dramatically and so fast had there been a minor price adjustment? Probably not. Would they have reacted differently if the power line had been permanently severed, as opposed to there being a three-month disruption? Probably.

This incident, and others like it, illustrates two points: (1) people react marginally to incremental price increases and *significantly* to *radical* price increases; and (2) they react differently if they perceive the price increase to be temporary versus permanent – the latter requires fundamental behavioural changes. Another lesson to take away from this particular example is that when the price of any necessity increases beyond a threshold level – we can debate what that may be – incremental adjustments are not sufficient; something more fundamental is needed.[2]

Incremental Price Adjustments versus Radical Price Adjustments

To illustrate what is meant by a radical departure from the status quo, let us consider another example. In May 2008 the government of Saudi Arabia announced that it would gradually reduce and eventually withdraw all agricultural subsidies for growing wheat in the kingdom, with the aim of a complete phase out by 2016. For a country that strove so hard for so many years to become self-sufficient in food, this was a radical policy change. The reason was not that the government could no longer afford the subsidies but that the kingdom had too little water; it decided that it made little sense to use so much of one scarce resource to grow a relatively low-value commodity, especially considering that other countries could do it so much more efficiently and cheaply.

If you think it was a stupid idea to grow wheat in the desert kingdom in the first place, let me point out that there are many examples of foolish use of a precious commodity because the price is not 'right.'

When Incremental Adjustments Are Not Enough

To give a sense of what is meant by incremental change, let us look at the Boeing 747, an aging airplane now superseded by the Airbus 380.

For years, Boeing executives were able to get more mileage out of the highly successful 747 by making incremental design changes. A little here, a little there, and the 747 remained the undisputed king of mass market, long-range air transportation for decades. However, there came a time when incremental improvements would no longer fly – literally. Boeing decided it was too costly and/or risky to redesign a substitute for the 747, allowing Airbus to do so.

The fate of the Airbus 380, of course, is still to be determined by how many orders the company ultimately receives, whether Boeing launches a late counteroffensive, and other factors, but there is no question that Boeing had run out of significant incremental improvements while Airbus decided to risk considerable investment in a radically different and larger design. The lesson one might take away from this example is that incremental fixes take one only so far, and this certainly applies to incremental price adjustments.

Price Elasticity

My next example, more relevant to the current debate in Ontario, comes from a personal experience in the water sector. One of my first assignments as a junior economist working with a multidisciplinary team of consultants in the late 1970s was to provide recommendations to the city council in Santa Fe, New Mexico, on pricing water. At the time, Public Service Company of New Mexico (PNM) ran an exclusive franchise managing the city's water system. I was the sole economist on a team consisting of hydrologists, geologists, civil engineers, and experts in piping, pumping, and water storage. Without getting into the details, the central facts of the case were that the city's water supplies were essentially finite and already stretched to the limit while the demand for water was continuing to increase steadily. It was a classic case of a supply-demand imbalance that had been known for a long time but not dealt with effectively.

Santa Fe is a lovely, affluent community, blessed with multiple amenities, ample sunshine, mild climate, beautiful scenery, good food, a thriving art community, opera – you name it. At the time of this study it was also relatively affordable and hence an attractive destination for retirees who enjoyed their backyard swimming pools, golf courses, and country clubs.

Over time, the city had gone from having low, fixed monthly water rates, without even bothering to meter consumption, to making timid increases in metered water rates. However, consumption had continued to grow. With demand outrunning available supplies, more drastic measures had to be considered, including water rationing, banning the use of sprinklers for watering lawns, and banning outright all nonessential gardening except for native, heat-tolerant plants.

Being young and naive, I recommended that prices be raised significantly. When people pointed out that the prior rate increases had not made much of a dent in consumption, I replied that it had to do with *price elasticity*, a term that none of them had heard before and which they ridiculed. Dropping economic jargon, I said that the prior price increases introduced by PNM had been incremental; the current dire conditions required *dramatic* price increases.[3] However, other members of the team pointed out that the city council would be voted out of office if it were to do such a thing. In that case, the only alternatives would be water rationing and/or changes in the city ordinance banning nonessential water use. I attempted, in vain, to argue that pricing would be a more efficient way to go: it would force consumers to rationalize how much water they would use, when, and for what purposes. In the end, it would be a more elegant solution and would obviate the need for rationing and resorting to a water police going around town to make sure people were not watering their lawns or washing their cars.[4]

Multiple lessons may be taken away from this particular example, including: when demand outstrips available supply, there is a problem; terms like *price elasticity* should not be used when talking to a city council; and suboptimal decisions are generally preferred by elected officials[5] – it is called *political expediency*.

Incidentally, Britain and South Africa are facing expected and current electricity shortages, respectively, having neglected to invest in sufficient new capacity.

Having It Too Good for Too Long

My next example, perhaps more relevant to Ontario, comes from another personal experience, in Tasmania, an island off the southern coast of Australia. This example is not about water but about electricity. In late 1980s I was hired as an 'unbiased' international expert to advise the minister of energy in Tasmania. Tasmania, which is blessed with ample rainfall and excellent hydro resources, had dammed a number of its wild rivers in the 1950s and 1960s, creating significant excess capacity. Water was literally spilling over the top of the reservoirs in wet periods, and nobody knew what to do with it. The engineers had clearly gotten ahead of the marketing people.

The solution was to attract energy-intensive industry to the remote island, certainly remote from the major international markets of the day. This required literally giveaway prices, but who was to say what was a giveaway price since water spilling over reservoirs did not have any value and there was no opportunity cost. The management of Hydro Tasmania, with the blessing of the politicians – who were essentially one and the same in the early days – decided that *any price* was better than zero.

Over the years, this policy worked, and a number of energy-intensive heavy industries were attracted to the island, most of them in energy-hungry metal-processing businesses. Raw materials were shipped to the island, and processed metals were shipped out to mainland Australia or exported overseas to be turned into gadgets and widgets. The finished products were sold to Tasmanians and others. There was little value added, and relatively few jobs were created. For a while, this arrangement made sense. Surplus electricity had little value in an island that could not export its excess capacity.[6] So long as the electricity generated some revenue – contributed to the margin – it was worth selling the power even at very low prices.

As the population of Tasmania grew over the years, the electricity demand outgrew available supplies. To make matters worse, at the time of my consultancy, Tasmania was in the midst of a multi-year dry spell, with reservoirs below their normal levels. The island's main thermal power plant, always intended as an emergency backup to the mostly hydro-based system, was now running at full throttle, using expensive diesel imported from the mainland and belching smoke and soot out of its stacks – a visible reminder that something was wrong.

When I arrived in Tasmania I learned first that twenty-two major industrial users were consuming roughly two-thirds of the generated power. Some of these users were sucking up the equivalent output of an entire power plant. More astonishing was the fact that very few people knew how little the major industrials (MIs, as they were called) were paying for the power. I quickly learned that the details of these long-term contracts were a tightly kept state secret. It took a lot of effort to find out the prices, and I was specifically warned not to disclose or discuss them publicly. The official line was that divulging the details of the confidential contracts would commercially harm the big users who were competing with one another. I suspected that there was another, even more compelling reason for not wanting to disclose the prices. The government and Hydro Tasmania officials would probably be embarrassed if the public learned just how little the MIs were paying for power – say, in the order of 2–3 cents/kWh – when everybody else was paying three to four times as much or more. The low rates could certainly have been justified when the contracts were initially negotiated, but during the interim years the MIs had bargained hard to keep them low, repeatedly threatening that they would shut down their operations and leave Tasmania, thereby causing massive unemployment and worse.

The options being considered by the minister and Hydro Tasmania were simple enough.

- They could build additional fossil-fuelled plants, with significantly higher costs to accommodate future growth – an option unpopular with everyone.
- Hydro Tasmania could build more dams on the remaining wild rivers – an option favoured by the engineers at the company but strongly opposed by environmentalists who did not want any major dams on the remaining scenic wild rivers.
- They could raise prices on the residential and commercial sectors to further promote conservation. [7]

There were a few other options that were not too attractive or practical. To my surprise, nobody was seriously thinking, and certainly not talking, about a fourth option, that of raising prices to the MIs, potentially significantly. In my view, this was the most sensible option. Moreover, in my mind, the issue was not really about rising prices but merely eliminating the significant historical subsidies that the MIs had enjoyed for too long and were not willing to acknowledge.

In a private conversation with the energy minister I was asked how likely it was that raising prices to the MIs would prompt them to shut down and leave. I told him that if prices were raised substantially, some of the marginal operators might close down. However, I pointed out that this would not necessarily be such a bad outcome. The MIs were electricity guzzlers who employed few people, were heavy polluters, and did not add much value to Tasmania's economy. They were essentially doing their dirty laundry because electricity was underpriced. If they left, the power could be put to better uses, creating higher paying jobs over time in other sectors of the economy including eco-tourism.

Like the prior example of expensive water being wasted in the Saudi desert in order to grow inexpensive wheat, here was a classical case of a scarce resource being overused and underappreciated because it was underpriced. In both cases, a precious commodity was being subsidized for reasons that were no longer justified.

Parallels and Lessons

At this point, one might ask why I present this extended discourse about price increases in Juneau, growing wheat in the desert, the aging 747 giving market share to the A380, water rationing in Santa Fe, and underpriced electricity in Tasmania. The reason is that there are important parallels in each case and possibly important lessons to be drawn for the policymakers in Ontario.

The 747 story was about running out of *incremental solutions*, at which point one needs to make fundamental changes that may include breaking with the past. One can surmise that the Province of Ontario is not alone in possibly having arrived at such a junction. Going forward, new, clean energy resources are likely to cost considerably more. We can view this as the proverbial question of whether the glass is half empty or half full. Some would say that electricity prices in the future may be significantly higher than those in the past, and that cannot be good for anyone. An alternative view may be that historically Ontario was blessed with low-cost hydroelectricity, but as higher-cost generation became necessary, industry, business, and consumers paid too little for power without even recognizing what a good deal they had. But the party is coming to an end.

The Santa Fe story was about *price elasticity*. I am not sure if the study's recommendations had anything to do with it, but if you visit Santa Fe today, you will notice that most homes and businesses have no lawns, relying instead on heat-tolerant native plants. The city is even more charming than it used to be and blends more beautifully into the surrounding dry, high plateau environment. Despite water shortages, it has managed to grow and prosper and has become even more affluent over the years. I can only surmise that water prices are much higher than they were in the past, and I suspect that people rationalize water usage, as predicted by economic theory. If you want to see green lawns in the desert that make no economic or ecological sense, you can still visit Las Vegas, Phoenix, and the rich Persian Gulf emirates.

The Tasmanian story, in my view, may be even more pertinent to

Ontario, and its lessons more pointed. It is about the need to adjust the price once it gets seriously out of line with its *relative value*. Historically, Ontario, the workhorse of Canada's economy, has effectively and successfully used its ample, low-cost electricity as a magnet to attract energy-intensive industry to the province. However, just as in Tasmania, the era of cheap, plentiful electricity has come to an end. Industry must bear the full burden of the replacement cost of new supplies, whether from the demand side or the supply side. Some businesses may decide that Ontario is not the right place to remain and grow; others may find that the higher electricity costs are worth the price because of the proximity to major markets and population centres. That ultimately is the right price.

Prices to Devices

As a young student in the late 1960s, I spent some time in London, England, in a bed and breakfast. The room had a small coin-operated gas heater. When it got cold, I would put a coin in the device and get some heat. Although I did not realize it at the time, it was a primitive but effective energy-control device. There was no need for a sophisticated value proposition. The coin clearly provided a certain amount of warmth.

If you visit the Electric Power Research Institute (EPRI) these days, you are likely to hear the term 'prices to devices.' The basic idea is that with advances in metering, two-way communication, and control technology – along with rapidly falling prices – it is not far fetched to think about a future where the price of electricity varies – not merely by time of use but by type of usage or device. Assuming that the devices have some intelligence built into them, can be programmed by customers, and can be remotely controlled by an intermediary or the utility, this opens interesting opportunities with significant implications for rational use of electricity.

Low-value, low-priority, high-flexibility uses of electricity (for example, the laundry) could be done when prices are below a given threshold. High-value, high-priority applications (say, use of computers, communication devices connected to the Internet, or data centres) will be served on demand at all times, perhaps at higher prices. Other devices with built-in storage would only operate during off-peak hours at lower rates. Plug-in hybrid electric vehicles (PHEVs) would be ideal candidates for such a pricing regime. Energy-hungry devices, such as air conditioners, may be cycled on and off, depending on preprogrammed commands. Consumers with roof-mounted solar cells, water heaters, or a small wind turbine in the backyard might sell power to the grid in certain hours and buy it back at other times, at different rates.

Such a scheme would allow intelligent devices to respond to intelligent prices, seamlessly and effortlessly, offering potentially significant cost savings even in a case where peak prices may be exorbitant. While there has been a lot of talk about such a future for a very long time, that future is not here yet. There are many reasons for that. The first obstacle is that people of middle age and older may simply be intimidated by the technical complexities. I am one of those people who cannot figure out how to set a programmable video recorder and who have trouble with all the features of a new mobile phone. The technology has to be integrated and made effortless if people like me are to adopt it.

That is not all. Since many utilities in the United States and, I assume, Canada are regulated, they must get permission before investing in new technology. I can tell you that the regulators in the United States – with a few notable recent exceptions – have not been consistent in supporting investments in so-called advanced metering infrastructure (AMI). In a 2006 survey, the Federal Energy Regulatory Commission (2006) found that a mere 6% of customers in the United States had smart meters on their premises. Sophisticated pricing schemes cannot be supported with dumb meters and an even dumber delivery grid.

Power and Promise of Price Signal

With rising gasoline prices, filling up the gas tank has become a painful experience. As drivers watch the dollar figures on the pump's display, they are made keenly aware of how much money is draining out of their pockets into the tank. For those with large cars, sport utility vehicles, or trucks and long distances to drive, this is an effective reminder to switch to smaller cars, drive less, carpool, take public transport, or telecommute.

For average electricity consumers, the bill may be painful when it finally arrives, but they have no idea how fast the dollars are adding up during the month.[8] This, many experts agree, is among the reasons that consumers may be using more electricity than they would if they knew how much it was costing them.

The fact that electricity-generation costs vary at different times of the day and across the seasons makes the problem even more acute. It also explains the sharp price peaks experienced by grid operators on hot summer days, something that is not well known to the average consumer.

Over the years, numerous studies have suggested that consumers would use less electricity if they knew how much it was costing them. The effect becomes more pronounced during peak demand periods when prices are significantly higher. The phenomenon is similar to studies that have documented that people walk more if they wear pedometers that count their steps; they eat less potato chips once the calories and the fat contents are clearly indicated; and they talk less when using public phones where the cost of the call is displayed on a monitor. The price signal is a powerful determinant of usage and certainly works as an effective deterrent to wasteful consumption.

In January 2008, the U.S. Department of Energy (DOE) released the results of a year-long experiment in the Seattle area, which concluded that when consumers are given the means to track and adjust their energy usage, power consumption declines by an average of 10%, and during peak demand periods by 15% (see box). The study, conducted by Pacific Northwest National Laboratory (PNNL), estimated that *smart grid* technology, if used nationwide, could save some US$120 billion in unneeded infrastructure investments, displacing the need for the equivalent of thirty large coal-fired power plants. Cost savings aside, that would be a large reduction in CO_2 emissions (*EEnergy Informer* 2008a). 'As demand for electricity continues to grow, smart grid technologies such as those demonstrated in the Olympic Peninsula area will play an important role in ensuring a continued delivery of safe and reliable power to all Americans,' said Kevin Kolevar, DOE's assistant secretary for electricity delivery and energy reliability.

Given such promising results, what is holding back widespread use of smart meters and programmable smart devices? There are four major hurdles: (1) there is a lack of *enabling technology*, the gadgets that enable the sorts of applications in the Seattle experiment; (2) more serious, the installation upstream of lots of sophisticated gadgets and downstream of a smart meter capable of two-way communication and remote control is not going to do any good unless all the parts of the system are *integrated* and work in unison, as was apparently done in the PNNL experiment; (3) the *behaviour* of large numbers of consumers has to be changed so that they can use technology that is complicated for most of us; and (4) most utilities have strong incentives to *sell more*, not less, electricity.

Box 8.1. Knowing Prices Influences Consumption

The PNNL study in **Olympic Peninsula**, near Seattle, included 112 homes equipped with smart meters, digital programmable thermostats, and computer controllers connected to water heaters and clothes dryers; they were supplied by **Invensys Controls**, and all connected through the Internet.

The sample homeowners could set their ideal home temperature on the Web and specify their level of tolerance for fluctuating electricity prices. Soon they began to manage their usage and control their electricity bills. 'I was astounded at times at the response we got from customers,' said **Robert Pratt**, a staff scientist at PNNL and the program director for the project. 'It shows that if you give people simple tools and an incentive, they will do this.'

Using technology developed by IBM, consumers could turn their thermostats and water heaters into bargaining tools, unaware of a sophisticated behind-the-scenes software. 'Your thermostat and your water heater are day-trading for you,' said **Ron Ambrosio** at the **IBM Watson Research Center**.

'We're not talking about traditional demand response where consumers have little or no control,' according to Mr. Pratt. 'We're talking about putting the power into the hands of the consumers, who can customize their energy use to save money or maximize comfort. They can check the financial implications of their decisions at any time, and adjust or override their settings whenever they choose.'

Referring to the Seattle experiment, Rick Nicholson, an energy technology analyst at IDC, a research firm, was quoted in a *New York Times* article as saying, 'What they did in Washington is a great proof of concept, but you're not likely to see this kind of technology widely used anytime soon.'[9] He was referring to the hurdles mentioned above. If the components of a smart grid/metering project are not effectively integrated, no amount of money or sophisticated gadgetry will achieve the conservation goal.

In March 2008, Xcel Energy Inc. announced that it would implement smart grid technology in Boulder, a smallish, upscale and environmentally conscious university town north of Denver. Starting with advanced meters and two-way communications for 25,000 homes – roughly half the city's existing homes – Xcel has set ambitious goals for the project (*EEnergy Informer* 2008c). Among the features of the new technology will be not only the ability to offer time-variable pricing but also the capability to offer different prices to different devices at different times. For example, plug loads and lighting can be charged at one rate – possibly a flat one – while air conditioning may be on a different and variable price; plug-in hybrid vehicles may be on an entirely different tariff so long as they are being recharged during off hours.

If everything works, and consumers can figure out how to program their devices – or can let the utility or a third party manage things for them – it would be a technical breakthrough. Eventually, it may even lead to a behavioural breakthrough as well. Over time, consumers may come to associate different prices at different times, as they do with their long-distance calls or mobile phones today.

Boulder's environmentally enlightened mayor, Shaun McGrath, believes that the new technology will help his equally eco-conscious citizens to move towards greater energy efficiency and heavier reliance on renewable energy and to make behavioural changes that would help the environment.

Why Boulder? The city is as affluent and green as any city that one could hope to find anywhere in the United States. In 2006, Boulder's citizens voted for a special tax of US$1 million per year to develop and implement a climate action plan that includes reducing energy consumption, eliminating reliance on coal-fired generation, and reducing the city's carbon footprint by 22% from 2006 levels.

Why Xcel? The company is under a state-mandated renewable portfolio standard (RPS) to increase the share of renewables from a mere 6% of generation today to 20% by 2020. With coal currently accounting for 57% of generation mix, it has a long way to go. Green Boulder offers a unique opportunity to test how far it can go with differentiated pricing, including charging a premium price for green electricity, something that appeals to affluent and educated consumers.

As already mentioned, U.S. investor-owned utilities routinely ask their regulators for permission to invest in advanced metering infrastructure before a nickel is spent. Xcel, however, has decided to proceed without seeking prior permission, confident that the project will be cost justified and will be approved ex post by the regulators. Mike Carson, Xcel's chief information officer, was quoted as saying that they were departing from the norm by seeking cost recovery after 'we have proven the benefits.'[10] It is a risk, and sets a bold precedent for the cautious, risk-averse industry that usually does not move a finger without first seeking assurance for cost recovery.

High-Value Electricity, Low-Value Use

I would like to conclude the discussion of what is the right price by citing one last example of how harmful dumb pricing can be. This is the American version of growing wheat in Saudi Arabia.

As many of you can recall, California experienced a serious electricity crisis in 2000–2001 with unreasonably high prices. It is a long story that I suspect many of you have heard before, and I will spare you the gory details.[11] At the height of the crisis, when California's independent system operator (CAISO) was buying all the power it could at any price just to keep the lights on, aluminum smelters in the Pacific Northwest decided that they could make more money by shutting their operations and selling their power allotment to California at a great premium. That story, I suspect, is well known.

Less known but equally interesting is the story of an Idaho utility that figured out a simple way to engage in a highly profitable form of demand response. It offered a decent reward to the potato farmers among its customers if they agreed not to use their electric pumps during the hot summer months. Quite a few agreed to take the money and forget about harvesting potatoes altogether. My recollection is that some 275 MW of peak load capacity was acquired at a modest cost. The utility was able to sell that to CAISO at a premium, pocketing the difference after paying the farmers. It was a classic win-win-win case.

What makes this demand response and the shutdown of the aluminum smelters interesting is that in both cases a high-cost, scarce commodity – peak hour electricity – was being used to produce a relatively low-value commodity – aluminum or potatoes, respectively. It does not take rocket science to see the fallacy of such a pricing policy, but I suspect there are many examples of dumb pricing such as this in Ontario and elsewhere if one starts looking for them. This is another way of saying that smart prices combined with smart devices and supported by smart policies is the way to go; it leads to smart, efficient, rational usage of electricity.

When the End Justifies the Means

Finally, I would like to say a few words about a case where a surcharge on price may be justified as a blunt policy instrument, beyond its traditional role as a conveyor of value. Such decisions require a value judgment, namely as to whether or not the end justifies the means.

Before the Second World War, the former shah of Iran decided that his backward, impoverished country needed a trans-Iranian railway that would connect the Persian Gulf ports to the north of the country bordering on Russia. It was a highly symbolic and expensive endeavour, and there was no easy way to pay for it. The shah decided to put a temporary surcharge on sugar, a staple that every Iranian uses to sweeten his or her tea. It was also a highly regressive tax, since it fell heavily on lower-income masses. Yet the people were generally supportive of the measure because they perceived the sugar tax as necessary, albeit painful, owing to the tangible results. The sugar tax was referred to by some as bitter-sweet.

The current debate about global climate change provides other examples of states, nations, and regions that are willing to pay a price today – in this case, to spare future generations from paying an even higher price. In the current context in Ontario, a political decision has been reached to phase out the province's coal-supplied generation over time. Assuming that the environmental end is supported by the majority of people, it justifies the means, which may include higher costs for substitute sources of energy.

Conclusions

I hope that the preceding discussion contributes a few useful thoughts for policymakers in Ontario. Following are some final considerations:

- Ontario is embarking on a carbon-light future with heavy dependence on energy conservation, renewable energy, nuclear energy, and smart pricing, which resembles the development of the Airbus 380 – and not the evolution of the aging Boeing 747.
- Businesses and industries in Ontario, like those in Tasmania, have had it too good for too long. The time has arrived for an adjustment in prices to reflect electricity's relative value and costs, which must include environmental externalities.

- Elected politicians must make difficult decisions and may have to convince their constituents that the end justifies the means.

Making broad recommendations, of course, is easy; implementing them is not, especially if they entail potentially significant price increases. Mindful of the political dimension, I am not advocating sudden price adjustments – and I am certainly not in favour of introducing price uncertainty or volatility. However, the message to be conveyed to consumers is the necessity of steady and stable price adjustments that will eventually reflect the higher costs of providing electricity services consistent with the new goals.

NOTES

1 For anecdotal evidence see, for example, EEnergy Informer (2008b, 2008d).
2 Economists have documented examples where relatively small increases in price produce relatively small decreases in demand. This is one reason for direct intervention in promoting energy conservation in the electric power sector. Utilities often intervene on behalf of consumers who only respond to significant price increases before engaging in energy conservation measures.
3 Experience in the electric power sector confirms the same: small price increases do not generally result in significant conservation or reduction in usage.
4 Nearly thirty years later, the exact same solution is being applied in Sydney, Australia, which is suffering from a serious multi-year drought.
5 For example, most economists favour a carbon tax to combat global climate change, but politicians prefer a cap-and-trade mechanism, even if it is more complicated to implement.
6 Tasmania is now connected to mainland Australia with an undersea cable.
7 Ironically, the MIs were paying so little for electricity that investing in energy conservation did not make much commercial sense.
8 In many countries, such as Britain, consumers are billed quarterly, making it even more difficult to relate consumption to the final bill.
9 *New York Times*, 10 January 2008.
10 *Wall Street Journal*, 13 March 2008.
11 For more details refer to Sweeney (2002).

REFERENCES

EEnergy Informer. 2008a. 'Power and promise of the price signal.' February.
– 2008b. 'Power sector construction costs surge.' April.
– 2008c. 'Xcel to showcase smart grid in Boulder.' July.
– 2008d. 'Why does everything cost more?' September.
Sweeney, James. 2002. *The California electricity crisis*. Stanford, CA: Hoover Press.
United States. Federal Energy Regulatory Commission (FERC). 2006. 'Assessment of demand response and advanced metering.' 8 August.
Wall Street Journal. 2008. 'Juneau residents brace for jump in electric bills.' 30 April.

9 The Effect of Price Elasticity, Metering, and Consumer Response on the Right Price

DEAN C. MOUNTAIN

My discussion will attempt to accomplish a couple of objectives. I will provide Ontario evidence to support some of the themes in the previous chapter, 'What is the 'Right' Price for Electricity in Ontario?' In addition, I will also provide extensions of some of the fundamental ideas pertaining to the 'right' price. Historically, Ontario electricity consumers have responded to electricity prices; some empirical supporting evidence will be provided. Recent Ontario experience with some market- and consumption-enabling technologies will be discussed. Furthermore, a refocus from price level to price structure is recommended. The extensions will also emphasize the importance of timely feedback of consumption and pricing information, the emerging enabling technologies for pricing of electricity, and relevant supporting market mechanisms.

The basic idea that price influences demand, whether the demand response be due to conservation or to load shifting, has not always been given the recognition or credit it deserves. Utilities and rate makers traditionally have been concerned about setting rates to recover costs and have paid little attention to price responsiveness. However, the evidence is clear. Ontario customers have responded to price. Table 9.1 provides a selected summary of estimates of Ontario elasticities, along with definitions of the various types of elasticities. The very recent study of Angevine and Hrytzak-Lieffers (2007) estimates real-time (hourly) elasticities to range between −0.10 and −0.14. Time-of-use industrial responsiveness has shown average hourly elasticities of substitution of 0.083, with a range from 0.0 to 0.380 (Cheng and Mountain 1993). In the residential sector, own-price elasticities for peak responsiveness (under a 3.9:1 peak–off-peak price ratio) have ranged from −0.095 in the summer to −0.142 in the winter (Mountain 1993; Mountain and Lawson

Table 9.1. Selected Elasticities for Ontario

Authors/Source	Rate	Elasticity		Remarks
		Average hourly elasticity of substitution[a]	Own-price elasticity[b]	
Angevine and Hrytzak-Lieffers (2007)	Real-time prices		[–0.10, –0.14][c]	Industrial Specific sectors had lower [0,–0.105]
Cheng and Mountain (1993)	Time-of-use	0.083 [0.00,0.380]	–0.050 [0,–0.298]	Industrial
Mountain (1993); Mountain and Lawson (1992, 1995)	Time-of-use		–0.095 (summer) –0.142 (winter)	Residential
Ham, Mountain, and Chan (1997)	Time-of-use		–0.078 (July) –0.085 (Jan)	Small commercial: offices and stores Compensated elasticities[d]

[a]Percentage impact on the ratio of peak to off-peak usage, in response to a 1percent change in the off-peak to peak price ratio.
[b]Percentage change of electricity demand in a particular time period, in response to a 1percent change in the price of electricity for that same time period.
[c][·,·]indicates range.
[d]Keeping the level of utility (for residential sector) or output/services (for industrial/commercial sector) constant, percentage change of electricity demand in a particular time period, in response to a 1% change in the price of electricity for that same time period.

1992, 1995). On the other hand, the time-of-use responsiveness in the commercial sector has not been consistent; nevertheless, for small commercial, offices, and stores their compensated price elasticities range from −0.078 in July to −0.085 in January (Ham, Mountain, and Chan 1997).

One recommendation in the chapter 'What is the 'Right' Price for Electricity in Ontario?' is that proper pricing of electricity should include the cost of environmental externalities. With electricity being a derived demand from other consumption and production choices, the connection to environmental externalities has been somewhat blurred. Most likely, for this to be successfully implemented an economy-wide commitment to include environmental externalities in the costing of all goods and commodities would be required.

With respect to price responsiveness, the above has illustrated that all demands are not the same. Some sectors are just not as responsive to particular price structures. We are too rigid when one price structure and set of rates is offered to all. Instead, rate making could be more sophisticated and allow for different prices for different customers.

While getting the right price and price structure is important, supportive and emerging enabling technologies can enhance the effectiveness of price. Part of delivering an effective price signal is providing instantaneous feedback on electricity usage. Traditionally, electricity has been one of the few commodities for which the bill is received one or two months after consumption. Surely, more immediate feedback on usage can only enhance the importance of the price signal. Recent pilot experience, where consumers have been provided instantaneous feedback using real-time monitors, has shown conservation impacts. Through radio signals, consumers instantly receive feedback from their meter on current kilowatts and total monthly kilowatt-hours along with the price per kilowatt-hour. Three Canadian pilots, Ontario (2004–5), British Columbia (2005–6) and Newfoundland and Labrador (2005–6), have shown significant, persistent load reductions. The average reductions were 6.5%, 2.5%, and 18.1% for Ontario (Mountain 2006), British Columbia, and Newfoundland and Labrador (Mountain 2008), respectively. Nevertheless, the diversity of loads and customers shows different reductions across various demographic groups (for example, lower responsiveness from seniors in Newfoundland and Labrador, and significantly larger reductions from electric water heating households in Ontario and Newfoundland and Labrador).

Among the next steps to occur regarding instantaneous feedback is to combine this with non-flat rates such as time-of-use prices, critical peak pricing, or some form of real-time pricing. Moreover, load control and/or conservation incentives may also be combined with instantaneous feedback. With the emergence of relatively inexpensive automated meter recording and instantaneous feedback, the door is now opening for both informative and imaginative pricing.

The development of appropriate market mechanisms to deliver the right-price signal is also crucial. For example, currently we have a spot market for large customers in Ontario. For a significant subset of these customers, spot pricing does not allow enough lead time for them to respond. However, one recent study (Deal and Mountain 2008) has shown that the development of a forward market, by contributing to market certainty and less volatility in overall prices, leads to more efficient responsiveness. The day-ahead market average elasticity of substitution across peak and off-peak consumption ranges between 0.307 when there is perfect certainty in the forward price relative to the spot price and 0.388 when there is 35% inaccuracy. Furthermore, the majority of customers prefer to make their consumption choices a day ahead versus in the day-at-hand spot market.

In conclusion, as noted in the previous chapter and from Ontario experience, price does matter and has an impact on overall usage and load shapes. However, just as we are encouraging more innovations on the technology front, we have to become more innovative on the pricing side. Part of this innovation requires acknowledging that not all customers are the same, and in turn appropriately designing a variety of price structures, each targeted at specific market segments with different elasticities and collectively inducing efficient price responsiveness. Let the customer choose, but have a rate design menu that encourages an overall efficient response while covering all costs. As well, the design of innovative prices can now accompany an array of conservation incentives and an array of enabling technologies that provide such features as instantaneous feedback, two-way communication involving the customer and the utility, and load control. Here we are at the frontier. While we have quite a bit of information on the impact of particular pricing policies and for some enabling technologies on their own, we have virtually no information on the impacts of the interaction of pricing and the emerging enabling technologies. Thus, an immediate priority is to undertake some pilots to assess which enabling devices in conjunction with specific rate structures and conservation incentives are appropriate and cost-effective for particular customer segments.

REFERENCES

Angevine, Gerry, and Dara Hrytzak-Lieffers. 2007. 'Ontario industrial electricity demand responsiveness to price.' The Fraser Institute Technical Paper, Energy Series. September.

Cheng, William, and Dean C. Mountain. 1993. 'Econometric study of the 1992 time-of-use impact of direct industrial customers.' Product Testing and Analysis Department and Economics and Forecasts Division, Ontario Hydro.

Deal, Ken, and Dean C. Mountain. 2008. 'Assessing whether firm day-ahead prices lead to changes in elasticity of demand and greater customer efficiencies.' *Independent Electricity Systems Operator*, May.

Ham, John, Dean C. Mountain, and M.W. Luke Chan. 1997. 'Time-of-use prices and electricity demand allowing for selection bias in experimental data.' *Rand Journal of Economics* 28: s113–s141.

Mountain, Dean C. 1993. 'An overall assessment of the responsiveness of households to time-of-use electricity rates: The Ontario experiment.' *Energy Studies Review* 5:190–203.

– 2006. 'The impact of real-time feedback on residential electricity consumption: The Hydro One pilot.' March.

– 2008. 'Real-time feedback and residential electricity consumption: British Columbia and Newfoundland and Labrador Pilots.' CEATI Report no. T041700-7013, June.

Mountain, Dean C., and Evelyn L. Lawson. 1992. 'A disaggregated nonhomothetic modeling of responsiveness to residential time-of-use electricity rates.' *International Economic Review*, February: 181–207.

– 1995. 'Some initial evidence of Canadian responsiveness to time-of-use electricity rates: Detailed daily and monthly analysis.' *Resource and Energy Economics* 17:189–212.

10 How Ontario Energy Institutions Set the Price for Electricity

GEORGE VEGH

Other chapters in this section (by Sioshansi and by Mountain) provide excellent examples of what one would expect to be intuitively correct: prices influence behaviour, and any attempt to change electricity usage should contain a price signal that is aimed at influencing behaviour. As a result, if it is desirable to have consumers use less electricity, electricity should have a higher price; if it is desirable to have consumers use electricity at off-peak times, then consumers should see time-differentiated prices; if it is desirable to have consumers use electricity produced with lower emissions, then an emissions component should be included in the price of electricity.

All of these goals – reducing and shifting the time of electricity consumption and reducing emissions in the production of electricity – are being pursued in Ontario. As indicated below, significant resources are being invested to pursue these goals. However, Ontario is not using electricity price signals to achieve them. Electricity prices remain subsidized (thus encouraging consumption, not conservation) and smoothed through regulatory manipulation (thus removing volatile price signals). In addition, in future, carbon emissions produced by coal would be effectively prohibited while carbon emissions produced by natural gas would be allowed. There is no price on carbon as such; there is a prohibition on emitting carbon through some forms of electricity generation and not others.

My comments explore these apparent paradoxes. First, I will elaborate on what constitutes a 'price' of electricity in Ontario with specific reference to the institutional make-up of those agencies that have the mandate of producing an electricity price. The agencies are the Independent Electricity System Operator (IESO), the Ontario Power

Authority, the Ontario Energy Board, and the Ontario Ministry of Energy and Infrastructure (the Ministry). The point of this section is to show that these agencies set the electricity price. Although supply and demand for electricity plays a role in this, it is a very small role. In other words, the agencies, not the market, set the price.

Second, I propose to demonstrate the ways in which electricity consumption and emissions are regulated in Ontario. I will focus on the steps that Ontario is taking to reduce consumption, to shift load to off-peak periods, and to reduce the carbon emissions resulting from electricity production. Although the agencies and the Ministry are putting considerable resources into these areas and using a variety of policy tools, price is not one of them.

Finally, I will suggest that the government's use of tools other than price is deliberate and not a symptom of a confused policy agenda. The government is not using electricity price as a policy tool for the simple reason that it is politically unpopular; indeed, the current institutional policy approach allows the government (directly and through its agencies) to obscure, rather than highlight, the cost of electricity. Although one can be critical of this approach from the perspective of efficiency and transparency, it has proven to be an effective way to bring about the results that the government and most people believe to be in the public interest: reducing the adverse environmental impact of the electricity system while providing low-cost electricity to consumers. As a result, before concluding that this is inappropriate, it would have to be demonstrated that the values of efficiency and transparency outweigh the value of effectiveness.

It is therefore arguable that the right price for electricity is the politically acceptable price and that environmental goals of electricity policy are more effectively achieved if they do *not* use price as a policy instrument.

What Is the Price of Electricity in Ontario?

In this section of the comment I will unpack the mechanisms of electricity pricing in Ontario by reference to the institutions that contribute to the components that add up to the price paid by customers for the electricity commodity.

The first component is the wholesale market price paid for electricity, which is the hourly market clearing price paid (initially at least) to electricity generators. Some context is helpful here.

The concept of an electricity commodity price is a fairly modern one. It accompanies the restructuring of the electricity market, which started in the late twentieth century across Europe and North America, and which came to Ontario in 2002. Prior to restructuring, a customer's cost of electricity was the price charged (usually at cost) for the full array of services provided by a vertically integrated electrical utility, including the production, transmission, and distribution of electricity through the electricity system. In Ontario, this was the cost incurred by Ontario Hydro and municipal electric utilities in producing and delivering power. With restructuring, the electricity commodity was unbundled from delivery. Putting a price on that commodity involved the creation of a market for that commodity.

The basic challenge for the development of an electricity market is to take the decisions that were previously made within an integrated utility and enable them to be made in a market transaction. Specifically, a generation division within a vertically integrated utility that also provides transmission and/or distribution services would provide generation when told to by the central command of a company. Requiring a generator to provide generation when customers want it requires an incentive. Furthermore, that incentive must operate within a physically integrated system, one that must keep supply and demand in constant balance. Sally Hunt (2002) expresses this basic requirement as follows: 'Whereas before, under command and control, the generators said, "Yes sir," now, under competition they effectively say, "What are you prepared to pay to compensate me for doing this?" But there is no time to negotiate about a price – *they must obey immediately knowing it is in their economic interest to do so*; they must know that they and the system operator are working under the same incentive-compatible rules that determine how they will be paid.'

In other words, the market price for an electricity commodity is the marginal cost of the incremental unit of electricity production. That marginal cost, in Ontario, is the hourly Ontario electricity price (HOEP). As a result, if one were to seek to uncover the price of electricity in Ontario, one would start with the HOEP. However, that would not go very far, because there are a number of government hedges and other actions that mitigate the impact of the HOEP. This mitigation affects the price signals that are sent to both suppliers and consumers. Suppliers are given signals about the relative value of investments in electricity production, and consumers are given signals about alternatives to electricity consumption.

There are two general categories of electricity suppliers: the government-owned generator, Ontario Power Generation; and private sector generators. Neither of these sources relies on electricity commodity price as a signal for investment. As discussed below, OPG is paid a regulated rate for the power produced by its baseload nuclear and hydro generation; for new generating facilities, OPG and private sector generators are funded through contracts with the Ontario Power Authority.

Half of OPG's generation capacity is made up of baseload nuclear and hydro assets. These assets receive a fixed annual price per kilowatt-hour that, by definition, does not have a relationship to the HOEP. The price received for electricity produced by these facilities is currently set by regulation, and the authority for setting that price is currently being transferred to the Ontario Electricity Board. The reason the government (and eventually the OEB) sets the price for this output is to ensure that the price is not the market price; in other words, one half of the price from OPG's assets is determined by an analysis that, as a fundamental premise, is unrelated to the market price.

Another portion (35%) of OPG's assets was, until 30 April 2009, subject to a revenue cap set by the government. The application of this revenue cap means that if OPG receives an average price of power from the wholesale market that provides revenue larger than the cap, the excess is returned to customers. In 2007 the OPG price limit was set at 4.7 cents/kWh. On 1 May 2008 the price was set at 4.8 cents/kWh. From 1 November 2006 to 31 October 2007, rebates of over $211 million were provided to customers.[1]

The price-regulated and revenue-capped components represent about 85% of the price of power supplied by OPG. The remaining 15% of OPG's power is set by HOEP. Thus, customers pay a largely regulated price for OPG's power.

The second source of investment in electricity facilities is the private sector. All private sector investment in Ontario since 2005 has received guaranteed public money; it has not been made on the basis of electricity price signals. The public money comes from the Ontario Power Authority, which has the obligation to purchase power, either at the direction of the government or following regulatory approval of the Integrated Power System Plan (IPSP) and Procurement Process[2] (which sets out the future generation-capacity and conservation goals for the province).

As of 31 July 2009, the OPA is currently managing 10,579 MW of generation capacity (Ontario's total generation capacity is 31,600 MW).

The OPA pays generators for power under long-term power contracts and collects its costs under these contracts from customers through the Global Adjustment Mechanism (GAM). Although the calculation of GAM is complex, it largely reflects the costs of the capacity payments that generators require to invest in Ontario that are not reflected in market prices. The GAM therefore moves in the opposite direction of HOEP; when HOEP is relatively high, GAM is relatively low, and vice versa.

The regulatory treatment of supply has direct impacts on the price signals sent to customers. Customers pay a mixed price that reflects the HOEP as adjusted by the regulated and capped prices paid to OPG and the contractual payments made by the OPA to private-sector generators. This mixed price leads to a mixed price signal, where the role of the HOEP is greatly diminished. In 2006 the Independent Electricity System Operator determined that the effect of this is a regulatory hedge of 81% of a customer's load. According to IESO (2006), 'for a $1 per MWh increase in the HOEP, consumers would receive an offsetting credit (reduced charge) through the GAM of $0.81 per MWh. In other words, a $1 increase in the HOEP would translate into a $0.19 per MWh increase in the overall Ontario consumer's energy bill.' On both a relative and an absolute basis, the role for HOEP has and will continue to decrease over time as all currently planned new supply is procured by the OPA. On a relative basis, as of September 2009, customers paid an annual kilowatt-hour commodity price of 3.15 cents/kWh plus a global adjustment of 2.63 cents/kWh.[3]

The Ontario Energy Board also plays a role in setting prices. As indicated, the OEB sets the price of OPG's baseload assets. As well, subject to government regulations, the OEB develops a formula for fixing prices for smaller customers through price forecasting and a quarterly smoothing mechanism. In other words, the commodity price paid by customers – which is already subject to the mitigation measures discussed above – has volatility further mitigated by a process of forecasting prices and reconciling the differences from the forecast.

The government's role in setting price is central to all of these decisions. As the owner of OPG, it has provided OPG with directions to develop supply that would otherwise be uneconomic; it sets regulations that the OEB must follow with respect to setting the price for OPG's baseload assets and setting the quarterly price adjustments; it also sets the revenue cap and provides directives to the OPA with respect to power procurements.[4]

The price of electricity in Ontario is therefore determined more by the decisions made by the Ministry of Energy and energy agencies than by forces of supply and demand. The result of these decisions is to subsidize prices in at least two important ways. First, in paying regulated and capped prices, consumers are not exposed to the marginal cost of electricity for a considerable portion of the electricity that is supplied to them. Second, even with respect to market prices, consumers are sheltered from price volatility through government hedges, such as GAM and OEB price smoothing.

It is clear that putting price fidelity into electricity regulation would require thorough reform of institutional design and mandate. In other words, if Ontario were to seek to achieve goals of conservation, load shifting, and reduced carbon emission, through price signals it would be necessary to fundamentally alter the approach to price that is currently taken. In the next section of this commentary I look at how these policy goals are being achieved through the electricity sector and why price has not been an important consideration. In the final section, I will ask whether there are good reasons to change the current approach.

Ontario's Current Approach to Conservation, Peak Shaving, and Emissions Reduction

It is arguable that a jurisdiction which does not allow for electricity price increases does not take seriously the issues of electricity conservation, peak shaving, and emissions reduction. In other words, higher electricity prices, particularly during peak hours, and some form of fee for carbon emissions arguably demonstrate a commitment to these causes; anything short of that is a failure of resolve to address these issues seriously. However, this is not the case in Ontario.

Considerable expenditures are being made on reducing electricity consumption. For example, for 2007 the Ontario Energy Board approved over $163 million in spending by electric distribution companies on conservation activities.[5] For the period 2008–10 the Ontario Power Authority forecast expenditures of approximately $1.3 billion on conservation and has proposed a plan to reduce net load between 2005 and 2025 through a mix of resource acquisition and improved efficiency standards (see 'Exhibit D-4-1,'*Integrated Power System Plan*).

With respect to reducing demand during peak periods in particular, the OEB estimated in 2005 that the cost of installing smart meters would be over $1 billion.[6] The installations have been largely carried out, but I at least have been unable to obtain a current estimate of the costs incurred to do this.[7]

With respect to reducing emissions from electricity production, the Province has committed to replacing coal-fired generation by 2014 through a combination of conservation, renewable resources, nuclear power, and natural gas. Coal is currently the cheapest form of energy; it will be replaced with energy that costs two to three times as much to produce. The approach of replacing coal will have direct environmental benefit. The greenhouse gas emissions produced by the electricity system will be reduced by 78% in 2027.

The point is that Ontario is investing considerably in reducing and shifting electricity use and in producing cleaner electricity. However, the Province is using a number of policy instruments to carry out this work, including utility rates, central agency budgets, and command-and-control measures; it is not using price. Rather, the approach is to socialize the costs of conservation and alternative supply sources and gradually collect them from electricity customers through various types of accounts. The result is that the customer pays but ultimately pays less (because many producers receive lower regulated and capped fees), while the higher fees paid to other producers are blended in with electricity prices, making all of the costs (including market prices) much less transparent. The net result is that the commodity price signal given to customers is inaccurate and obscure.

Ontario is not using price, because changing electricity prices – more specifically, *increasing* electricity prices – is politically very unpopular. It should be borne in mind that at the same time the government was designing its policies respecting conservation and coal, it was also recovering from the political backlash that had faced a previous government when it sought to use transparent electricity price signals through primary reliance on HOEP as setting price signals for both consumers and suppliers. It is not necessary to repeat the story of the rise and fall of competitive electricity models in Ontario. It should suffice to say that the electricity market was opened in May 2002; the market price for electricity increased in the summer of that year (though it was still lower than in neighbouring jurisdictions); and the government went back to fixing prices in November 2002.[8] The point here is not that the policy adopted in November 2002 was a good one;[9] the point is that the Ontario electorate responds negatively to electricity price increases and volatility. The fact that this reaction may be somewhat irrational does not make it any less real.

Should Ontario Be Using Price Signals?

The foregoing discussion sought to demonstrate two points: (1) Ontario electricity regulation does not use price as a policy instrument, and (2) Ontario is making a substantial investment in pursuit of the electricity policy goals of conservation, peak shifting, and emissions reductions, using non-price policy instruments.

The question for consideration here is whether or not Ontario should change its approach to using price signals to either reinforce or replace its current approach to addressing the environmental impacts of electricity.

This is a complex issue because it involves first determining what it is that these environmental policies are seeking to achieve; in other words, it is necessary to first identify the problem that the policy tool of a price signal is trying to achieve. If the problem is largely one of economic inefficiency, then the price signal would be helpful. Specifically, if the environmental concern of electricity regulation were about preventing *inefficient* environmental impacts, then the price signal would both define which impacts are inefficient and provide direction on how to achieve efficiency. Thus, if the conservation goal were aimed at avoiding inefficient electricity consumption, inefficient use of peaking electricity, or inefficient production of carbon, then pricing electricity by reference to its economic alternatives would be the appropriate way to achieve these ends.

It is not clear, however, that the problem is one of economic efficiency. The government is seeking to use the electricity system to achieve environmental benefits whether or not the outcome is economically efficient. Further, the most striking aspect of the government's conservation and clean supply agenda is that, although it has imposed burdens on taxpayers and ratepayers, this has not resulted in any significant political backlash – perhaps because the burdens are virtually invisible in the opaque landscape of government revenue. I believe that the lack of a backlash is largely due to the use of non-transparent pricing policy; the pure commodity price has been kept low, but increases have been levied in a number of other charges. Lately it seems that the increased political cost of using price signals as a policy instrument outweighs the benefits of price fidelity.

It is not clear whether or not such an approach to pricing is enduring. However, in order for the approach to change, it will be necessary to identify a political imperative, not simply an efficiency imperative. This will require a constituency in favour of pricing signals. Although this constituency has not yet emerged, one would expect that there would be prospects for a coalition of environmental groups (who seek higher price signals to support conservation and make renewable power more competitive), industrial consumers (who seek the ability to hedge their exposure, which is almost impossible in the current environment), and generators (who will receive more revenue under a market price regime). This type of constituency will have to overcome many natural forms of opposition; environmentalists, consumer groups, and generators have conflicting interests in other areas of energy policy, and bringing them together could be difficult. Unless and until such a constituency does come together, the government's approach is unlikely to change.

NOTES

1 See 'OPG rebate: May 1, 2006 to April 30, 2009' at http://www.ieso.ca/imoweb/b100/b100_ONPA.asp.
2 The Ontario Energy Board's review of the OPA's application for approval of the Integrated Power System Plan and Procurement Process was adjourned in October 2008 and remains adjourned at the time of writing.
3 See http://www.ieso.ca/imoweb/siteShared/electricity_bill.asp?sid=bi.
4 There is overlap in these areas. Although the OPA contracts primarily with private investors, it has also been directed by the government to contract with OPG to pay for the costs of developing hydro resources.
5 See 'Conservation and demand management: Reports, plans, and applications' on the OEB website, http://www.oeb.gov.on.ca/OEB/Industry+Relations/OEB+Key+Initiatives/Conservation+and+Demand+Management+(CDM)/CDM+Reports+Plans+and+Applications.
6 See http://www.oeb.gov.on.ca/documents/cases/RP-2004-0196/sm_draftiImplementationplan_091104.pdf.
7 The best description of the current types of demand-response programs in Ontario is found in the OEB Market Surveillance Panel's *Monitoring report on the IESO-administered electricity markets for the period from May 2006 to October 2006*, available at http://www.oeb.gov.on.ca/documents/msp/msp_report_final_20061222.pdf.
8 For a good discussion, see Trebilcock and Hrab (2006).
9 I have argued elsewhere that it was a bad policy, in Vegh (2002).

REFERENCES

Hunt, Sally. 2002. *Making competition work in electricity*. New York: John Wiley & Sons, 34–5.

Independent Electricity System Operator (IESO). 2006. 'The role of fixed price contracts and the global adjustment in mitigating price changes to consumers.' 19 October.

Trebilcock, Michael J., and Roy Hrab. 2006. 'Electricity restructuring in Canada.' In *Electricity market reform: An international perspective*, ed. F. Sioshansi and W. Pfaffenberger, 419–41. Oxford: Elsevier.

Vegh, George. 2002. 'The price of electricity is not too high.' *Globe and Mail*. 12 December.

PART FOUR

Policy Tools for Increasing End-Use Electricity Efficiency

11 Introduction

BRYAN W. KARNEY

Humans have collectively enjoyed remarkable success in increasing the production and use of energy for human ends. Although historically a well-trained and organized set of workers using their own strength could provide enough motive force to gradually transform landscapes and to build impressive monuments, such achievements pale compared to the power that modern society routinely employs. Even the use of animal power, such as horses, is dwarfed by modern machines. A car engine may be equivalent to several hundred horses, and a jet engine to tens of thousands, but to evaluate a major electrical generation facility as equivalent to millions of horses is surely to make the comparison practically meaningless. Even a mere litre of gasoline, an energy unit so familiar that it is readily dismissed for its insignificance and currently priced at less than a dollar, represents the energy equivalent of roughly ten days of manual labour for an average person.

These advances in harnessing energy, however, have been bought at great cost to the natural system, bringing many unintended consequences. Our insatiable quest for energy consumes mountains of coal, causes whole rivers to be re-routed, taints the air and water, and is altering the composition of the global atmosphere. These concerns raise profound questions of sustainability and of whether we can learn to use energy in a way that is more conservative in its consequences. Such issues are at the heart of part 4 of this book.

The lifestyle gains associated with energy use are often adopted more easily than they are relinquished. Thus, we must consider whether the same lifestyle can be purchased at a lower energy and environmental cost. That is, can efficiency measures be introduced that simultaneously conserve energy and preserve lifestyle? Framed in this way, the key questions are, how efficient is our energy use and how do we reduce the environmental burden that is associated with a given measure of energy service (heating, cooking, cleaning, moving) and not with the energy use per se?

In the first chapter of this part, Loren Lutzenhiser considers the policy tools that might effectively promote energy conservation and enhance energy-use efficiency. With great insight and an admirable sense of history, Lutzenhiser raises these issues both generally and in the context of the California experience. He puts energy use not only into the context of new devices and more efficient components but squarely in the domain of human behaviour.

It is easy to make assumptions about what people would or would not do, but how do people actually respond to various policy inducements, and how might long-term behaviour be changed? Although Lutzenhiser argues that human behaviour is complex and not simply captured by glib assumptions of a self-serving rationality, his chapter raises a significant measure of hope. Under quite specific circumstances that are clearly articulated, people are shown to be capable of greatly exceeding conservation expectations.

Yet, an interesting and ironic background question cannot be escaped. If technologies become more efficient, how will people respond to the resources that are thereby freed up for other uses? Certainly it is unlikely that the environment will be the sole beneficiary of efficiency gains, meaning that the actual energy usage decreases will often not be in lockstep with efficiency improvements. Moreover, if a given activity effectively costs less because its energy use is smaller, is it possible that additional energy-using activities will be then adopted? Indeed, is it even possible that conservation measures can lead to behaviour that actually increases energy use? This phenomenon is called *rebound*. Rebound is sometimes dismissed with the argument that it makes no sense: after all, how could one advocate an energy-inefficiency technology? Yet, rebound is sometimes used by others to dismiss outright any attempt to improve energy-use efficiency. In this context, the key challenge is to understand not what people *might* do, but what they *will* do, based on what they actually have done.

In chapter 13, Steve Sorrell provides a magnificent summary of the experience with rebound, what factors influence it, how it arises, and how it actually plays out in a variety of contexts. With patience and a thorough discussion of available data and sources, this chapter beautifully frames the policy implications of the rebound question. Significantly, but perhaps not too surprisingly, the truth is closer to the middle than extreme views would suggest. Thus, Sorrell argues that rebound does occur and that the policy measures that fail to account for it are doomed to at best disappointment and at worst failure. However, significantly, Sorrell concludes that policies to improve energy-use efficiency remain vital for reducing the consequences of energy use.

In chapter 14, Mark Jaccard briefly summarizes the positions of Lutzenhiser and Sorrell and suggests how he would use the research on which they report to assess the appropriate roles for price and other policies in achieving energy and environmental goals.

12 The Evolution of Electricity Efficiency Policy, the Importance of Behaviour, and Implications for Climate Change Intervention

LOREN LUTZENHISER

Significant improvements in the efficiency of energy use will be integral to global efforts to reduce greenhouse gas emissions. Parts of the world have had some success in securing energy efficiency (EE) as a policy goal at the national level, and some regions of North America and Europe have accumulated considerable EE experience. Future policies, programs, and initiatives should be able to benefit from what we have learned over the past thirty years. Nevertheless, the EE policy paradigm, which primarily encourages the proliferation of modified hardware – appliances, buildings, infrastructure – in support of gradual and relatively modest aggregate gains in energy efficiency, will be insufficient to stave off the immediate threat posed by global warming.

The chapter is about energy efficiency and policy, past, present, and future. It pays special attention to the role of human behaviour in energy use and conservation and to the policy implications of that relationship. To make policy is to attempt to change human behaviour. In the EE arena, some simple examples come to mind. The first has been to legislate increasingly stringent appliance and energy use standards, resulting in continual modifications to home and office hardware. Here, the choices and actions of regulators, designers, and manufacturers changed the hardware. A second has been to encourage and/or reward particular EE choices – such as the purchase of more efficient refrigerators – which relies upon distributors, retailers, and consumers behaving differently. A third has been to inform and urge the more efficient use of existing energy-using devices (for example, dishwashers, lights, air conditioners), asking people to change their practices. In each case, reduced consumption depends on the ability of policy to successfully impact upon behaviour.

While all EE policies have attempted to change behaviour, they have largely been designed to compel producers to create more efficient products and to induce consumers to purchase them. In so doing, they have demonstrated both a simplistic understanding of how people make decisions and a disregard for the way people actually interact with a technology once it has been adopted. Unlike other policy areas where behaviour has long been a central focus (public health, education, and crime prevention come to mind), EE policy discourse, program design, service delivery, and evaluation have been almost exclusively *device-centred*. We know that human behaviour and choice are crucial elements in the lifecycle of energy-using devices, but where choices have been considered to any degree, they have been viewed narrowly in economic terms, as contests between calculations of costs and benefits with optimal rational solutions. Decision makers, as rational actors, are assumed to be informed or at least able and willing to use information provided by utility companies, government agencies, and non-profit advocates to make appropriate, efficient hardware choices.

Research and program experience have repeatedly demonstrated that such choices are rarely economic, rational, or informed, as I discuss below in detail for California households. However, these findings have had little effect on the models, methods, and policy frameworks routinely used in the EE world. No well-accepted alternatives from the social and policy sciences have been offered to take into account the complexity of human behaviour. Therefore, policy change in EE has largely meant refining and making more complex approaches that are grounded in the same basic assumptions of rational choice. These are joined with a perspective that sees the world populated with devices and physical environmental conditions. The result is a deceptively solid physical-technical-economic model (PTEM) of energy use (Lutzenhiser 1993). Nascent efforts to take human action seriously in EE thinking tend to take the form of psychological amendments to the PTEM.[1]

This chapter takes a critical look at these issues. It offers a sober assessment of the potential of EE policy and program experience to support dramatic expansion of EE activities with significant improvements in societal energy efficiency. My primary interest is in the electric power sector, and particularly residential energy use, the target of significant EE activity over the past three decades. My particular focus is on the evolution of electricity efficiency policy and intervention approaches in California and the U.S. Pacific Northwest. While significant EE development has certainly taken place elsewhere, the scale and breadth of innovation has been particularly impressive on the west coast and influential at the global level. So it would seem reasonable to expect that region to offer a wealth of social learning from which ideas and approaches could be applied elsewhere.

The balance of this chapter is divided into three parts. First, I review the evolution of EE policy from the 1970s through the 2001–02 California electricity supply crisis, sketching out a context for the present state of policy and programming. Second, I consider the surprising roles of both consumer choice and conservation behaviour in the crisis and how that experience may have affected California's consumers in the longer term. Finally, I examine the lessons learned from more than thirty years of EE policy experience and assess the fit of EE policies to ambitious climate change goals. This involves considering the potential contributions of the social sciences to a climate change policy discourse that takes human action seriously, as well as considering both promising policy developments and a number of impediments to the rapid expansion of EE activities.

Evolution of Energy Efficiency Innovation in California and the Pacific Northwest

California and the Pacific Northwest might be the least likely of places to be focused on saving energy. The region is home to an abundance of natural resources, including coal, oil, natural gas, and enormous watersheds that power vast hydroelectric systems. The energy systems of California, Washington, and Oregon are closely interconnected with one another as well as with adjacent states (for example, Nevada, Utah, Idaho, and Montana) and British Columbia, jurisdictions that are also energy rich. Moreover, electricity prices in the Pacific Northwest states and BC have historically been among the lowest in North America, permitting a widespread use of electric space heating and water heating that is unheard of in many other locales.

In the post–Second World War period the region was an exuberant and expansive place. However, along with other growing, sprawling, and rapidly developing parts of North America, by the 1970s it clearly showed that the demand for energy would sooner, rather than later, outstrip supply. A flood of plans for new coal mines, natural gas wells, power plants, transmission lines, pipelines, and nuclear power stations was proposed as a solution. Regulated electric utilities were allowed to collect a rate of return on investment that was negotiated with state and provincial utility commissions. Opportunities to expand investment in new sources of supply and generation were also opportunities to increase returns to shareholders. Government utilities, while not rewarded by profits, nonetheless saw looming shortages and also moved to invest in new energy sources.

This 1970s wave of planning, borrowing, and building crashed headlong into new social realities. They included energy crises at both the beginning and the end of the decade, growing environmental consciousness, the rise of an organized environmental movement, and the expansion of citizen activism aimed at 'big' targets (for example, big oil, big business, big technology, big government). Nuclear power proposals in the region were troubling to many, particularly after the meltdown of the reactor core at Three Mile Island in 1979, which lent credence to longstanding concerns about safety.

In the northwest a group of public utilities experienced massive cost overruns and delays in constructing five nuclear plants in the early 1970s. The plan began to unravel in 1974 with the withdrawal of the City of Seattle from its commitment to nuclear power plants and culminated in large-scale bond defaults, lawsuits, and ultimately the completion of only one plant. Concern about energy grew, influenced by the oil embargo and energy crisis of 1973, as well as by water-shortage years that reduced the availability of hydroelectric power generated by a system of huge federal dams; they had been built during the 1930s as public works projects and were administered by a federal agency, the Bonneville Power Administration (BPA).

Concrete policy actions to control the willy-nilly expansion of supply and to minimize the costs of overbuilding took several forms. In California, the *Warren-Alquist State Energy Resources Conservation and Development Act* in 1974 created a state agency charged with assessing and forecasting the rates of growth in demand for electricity and natural gas, in order to provide a standard against which to judge whether or not power plant proposals should receive state siting approval. Following several years of increasingly acrimonious conflict, the U.S. Congress passed the *Pacific Northwest Electric Power Planning and Conservation Act* in 1980. It created a four-state agency whose board would be appointed by the governors of Washington, Oregon, Idaho, and Montana. The staff of this Northwest Power and Conservation Council (NPCC) was made responsible for independent forecasting and planning, similar to what was being done in California by the new California Energy Commission (CEC).

The Northwest Power Act, however, had two innovative features. First, it included environmental protections, mandating that the BPA work to protect migratory salmon populations in the Columbia River system. This mandate has dominated the concerns of the NPCC since its creation. Second, it asserted that energy efficiency was the preferred source of new supply and should be pursued as a least-cost alternative to the building of new generating facilities of any type. In both California and the Northwest then, a priority for the reduced use of energy, freeing up savings for new sales – that is, *energy efficiency* – has been a central feature of electricity-related regulatory policy for nearly thirty years now.

Institutionalizing Energy Efficiency Policy and Programs

A wide array of programs and energy-savings evaluation activities has been implemented in the region. Numerous institutions that plan, deliver, assess, research, and support energy efficiency activities have come into existence. These include state energy agencies, national research laboratories, state research and development programs, utility-sponsored EE support and acquisition programs, university research centres, and public-private consortia and coordinating organizations. This 'energy efficiency industry' also includes a large number of private engineering, consulting, analysis, and EE program delivery firms working in the region and employing a significant number of EE specialists of one sort or another – from engineers and policy experts to marketing, program management, and modelling practitioners.

The range of policy approaches is similarly broad, with a varied set of instruments deployed to reach different sorts of targets. Still, all function in legalized regulatory contexts where technical issues dominate, solutions are highly engineered, and economic analysis strongly influences thinking about alternatives, policy choices, and program delivery.

The instruments and activities include the following:

- Laws, regulations, building codes, and hardware performance standards.
- Modelling, forecasting, and planning processes.
- Public information (ranging from formal hearings to consumer websites and the delivery of EE information with monthly utility bills).
- Technology research and development (for example, related to buildings, devices, and control systems).

- Technical assistance, information, and education (from lighting and systems design assistance to the provision of data on comparative efficiencies of equipment, calculators for costs and paybacks, technical training courses, and programs in schools).
- Payments, incentives, rebates, grants, and tax credits (direct infusion of money, from government and ratepayer sources, to encourage, induce and/or subsidize EE hardware installation, and even sometimes to encourage behavioural conservation, as in the case of the California 20/20 program during the 2001 electricity crisis, which reduced power bills by 20% if a savings of at least 20% was observed).
- Free hardware, installation, and maintenance (particularly in the case of low-income energy users, and sometimes in the form of public light-bulb distribution or free audits and tune-ups).
- Rates, taxes, and fees (the 'stick' to go with, or instead of, the 'carrots;' for example, higher prices for larger volumes of energy use or for consumption during high-demand time periods).

All of these demand side management (DSM) activities are thought to provide energy savings as well as non-energy benefits, such as improvements in quality of life and/or environmental conditions. However, it is important to note that the primary policy justification for these activities and expenditures is the logic of 'avoided cost.' Demand side management delivers energy to the system at costs that have, in most time periods, been lower (often much lower) than the costs of new generation in the forms of coal, natural gas, petroleum, or nuclear power plants. These 'negawatts' (Lovins 1989) were once used inefficiently but, after EE upgrades, become available for sale to new customers or to old customers for new energy uses.

This supply is created out of savings from many sources, so the cost of an individual negawatt is quite variable. Some are inexpensive and pay for themselves, with costs recouped by savings on energy bills in a very short time. Others produce smaller savings at higher costs. All can be ranked by cost and savings on an EE 'supply curve' (Meier, Rosenfeld, and Wright 1982) that can be used to compare technologies and prioritize programs, expenditures, and investments (see figure 12.1).

The avoided costs of repairing or mitigating the environmental harm that can result from conventional generation technologies – salmon population declines, forest damage from acid rain, health costs from air pollution, and the myriad deleterious effects of global warming – can

Figure 12.1. Energy Efficiency Supply Curve

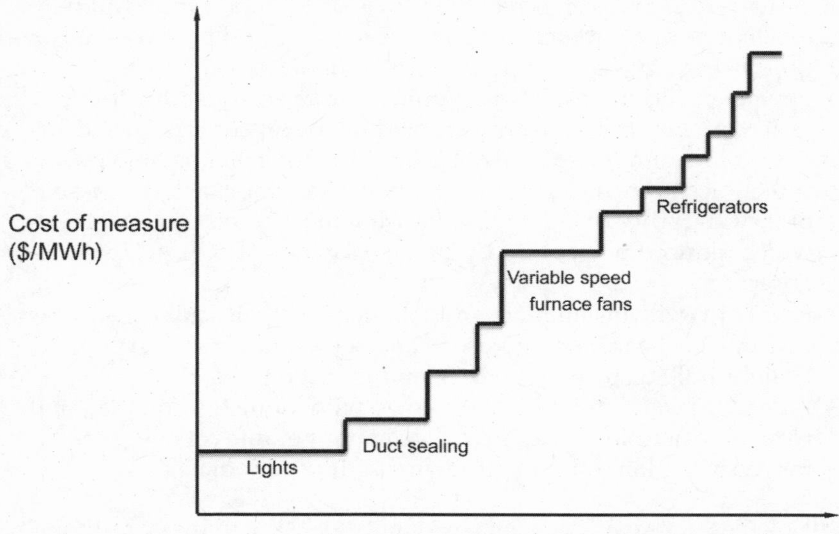

be included in least-cost-planning calculations, making efficiency even more cost-effective as an energy resource. As a result, EE has enjoyed a privileged position in electricity resource acquisition policy, as the first option in Northwest power planning and the top of the loading order in California utility regulation.

The cheapest points on the supply curve are sometimes called *the low-hanging fruit*, and some of this has been 'picked' over the past thirty years. But much is thought to remain, for several reasons. First, the goal of EE policy has been to *slow* system expansion, and in most time periods fairly modest efficiency gains were all that were needed (and all that were funded) in order to avoid construction of new power plants. Modest support leads to modest results. Second, though policy is designed according to how economically rational firms and economically rational consumers ought to behave, real world choices do not seem to reflect the hypothetical imperative for easy savings, low costs, and minimized risks.[2] Finally, declining fuel costs (particularly natural gas) and new electricity generation technologies (for example, gas combustion turbines) contributed eventually to declining levels of commitment to efficiency. Therefore, while interest in EE has been lasting and subject to policy and programmatic innovation, the results have yet to meet the potential.

Markets and Meltdown

The 1970s saw the birth of EE on the west coast.[3] The 1980s and early 1990s were the years of institutionalization and elaboration of EE policies and practices. By the mid-1990s, however, the energy landscape had changed. Falling natural gas prices and the development of relatively inexpensive and 'down-scalable' turbine technology suddenly made producing more energy look inexpensive again, and large industrial energy users led a growing call for the deregulation or restructuring of the electricity grid that would provide them access to lower-cost electricity. Large firms threatened to generate their own electricity; speaking at an energy conference, one executive said, 'We've been making steam for a long time, so it's no mystery to us how to operate a gas turbine.'[4] Growing interest led to speculation about the more widespread benefits of a freer market for electricity, where new entrants would compete with utilities and possibly lower electricity prices overall. A sentiment commonly stated at energy meetings during the mid-1990s was, 'It isn't a question of *if* deregulation will happen, but *when*.'[5]

The consequences of deregulation for EE were uncertain, but three possible scenarios were imagined. In one, the cost advantages of investment in more efficient technologies would be recognized by savvy consumers in a deregulated environment. Markets would actually increase the uptake of the low-hanging fruit, and governmental intervention would no longer be necessary. In a second scenario, natural-gas-generated electricity and intense price competition in newly created electricity markets would eliminate the cost-competitiveness of EE, driving it from both the policy arena and the market. The third vision, which had considerable influence in California and the Northwest, saw policy and the market working together. While much-reduced EE investments could be accepted in a deregulated energy economy, the benefits of EE programs and attendant social learning were too valuable to sacrifice completely to the market. Strategically targeted interventions into particular markets were thought to be wiser, and market transformation (MT) was born. This approach sought to alter the mix of energy-using technologies available in the market, through bulk purchasing, incentive payments to manufacturers, subsidies to retailers, and other interventions in supply chains at points upstream from end-users.[6]

Deregulation did happen in the American west, but not everywhere and not in one particular way. The regulated system was hardly changed in some states (including Idaho and Washington), while Oregon went for a partial restructuring that provided for very large ratepayer-funded EE investments. Montana and California opted for radical deregulation, and the results were disastrous. Montana's primary utility suffered a collapse that proved tragic for shareholders, employees, communities, and consumers. The California case, however, is instructive in four important ways.

The California deregulation or 'regulatory restructuring' plan came about with considerable deliberation among a host of interested parties, with ironic results. The legislators, regulators, utilities, large energy consumers, environmental organizations, cities, ratepayer advocates, and large universities involved in the discussion finally agreed to divest half of the regulated utilities' electric generation assets (power plants) while opening the markets to some forms of wholesale competition. The utilities would be able to charge higher initial rates to cover the costs of retiring the 'stranded assets' that had been built before deregulation had been imagined and at costs that did not reflect the subsequent declining prices of natural gas. After those investments had been recovered, retail prices would be allowed to reflect presumably lower market prices.[7] At the same time, it was recognized that the utilities routinely did more than simply produce and sell power to consumers; as a result of past regulatory decisions, they also had come to provide a variety of public goods. These included investments in environmental protection and impact mitigation, low-income subsidies and home weatherization services, a variety of energy efficiency programs, and support for collaborative research and development on new technologies. Rather than see all of these investments disappear under deregulation, the California legislation provided funding for various 'public purposes' through a small, uniform tax or fee on electricity sales.[8] This innovation has been emulated in a number of other states, with or without accompanying degrees of deregulation. Therefore, as it turned out, a plan that aimed to reduce the role of government in the energy system actually resulted in the institutionalization (and sometimes expansion) of public goods activities both inside the utilities and across various government agencies.

The second lesson we can learn from California's deregulation experience is that the most carefully crafted plans by the most impressive array of experts cannot anticipate all eventualities. The California process ran smoothly enough for several years, from the passage of legislation in 1996 until the summer of 2000, when the San Diego area was the first area to be exposed to market prices. By the fall of 2000, unpredictable and extreme wholesale prices (passed on to retail consumers) had hit the other parts of the state. Escalating costs were made worse by electricity shortages (later shown to be often manipulated by wholesale generators), the ultimate bankruptcy of one of the state's largest investor-owned utilities, and the state government's intervention as the last-resort purchaser of power on the market for California consumers.

Third, an unanticipated and quite ironic policy consequence of the California deregulation experience was the virtual abandonment of market transformation in 2001 and a wholehearted return to EE supply acquisition by regulators, who subsequently devised new ways to allow utilities to profit from selling less electricity.[9] Recently, however, interest in MT has returned to the energy scene in California, and a mixture of EE and MT policies and interventions persists in the Pacific Northwest.[10]

Finally, the California case proved that consumers could be persuaded to respond positively in times of crisis. With all options exhausted, California was left pleading with consumers to voluntarily make choices that would conserve electricity. Those consumers surprised everyone by changing their behaviours and saving the system from further shortages, blackouts, and disruptions to the economy and society.

Consumers and the California Crisis Response

By early 2001 it had become apparent that shortages and high prices in the electricity markets (upon which the state depended for half of its power in the deregulated system) had compromised the ability of the California utilities to supply electricity. The utilities and the grid operator developed a strategy of rolling blackouts. When the demand for electricity exceeded the available supply (or supply available at an acceptable price), intentional short-term outages were rotated among clusters of consumers, neighbourhood by neighbourhood. Customers were notified of their outage assignments, but uncertainty was high, news coverage was intense, and calls for government action were strong.

Meanwhile, the State of California was forced into the role of electricity buyer for the major utilities. The governor and legislators seriously considered seizing the assets of the utilities and even those of the third-party generators. Nearly US$1 billion was appropriated to fund energy-saving efforts, and the EE policy arsenal of government and utilities was deployed rapidly and aggressively. So too was a novel communications effort dubbed 'Flex Your Power.' The state-wide campaign asked consumers – via print, radio, and television ads that stressed the importance of even the smallest gestures – to voluntarily conserve electricity. Consumers were asked to be conscious of their energy use: to turn off unneeded lights and appliances and to shift the time of laundering, cooking, bathing, and using air conditioners away from the late-afternoon system peak.

All of these efforts resulted in stabilized supply levels and reductions in demand levels that prevented additional rolling blackouts and obviated the crisis. The state-wide goal of a 5,000-megawatt peak reduction – initially thought to be ambitious – was exceeded by 20%. Energy loads in businesses, government offices, and homes were lowered across the state. The story of how this all took place is instructive for policy discussions about the respective strengths and limitations of energy efficiency and behaviour-based conservation.

Lifestyle and Thin Ice

The voluntary behavioural change on the part of consumers in California was actually quite unexpected. The EE policy toolkit available for deployment in 2001 was limited, having evolved as a means of securing small amounts of negawatts over long periods of time. It was end-use device centered and assumed an economic investment logic and cost/benefit calculus on the part of households and organizations. It could only try to induce the purchase of new appliances, motors, lighting fixtures, or industrial equipment; changes in behaviour – conservation as opposed to hardware efficiency – were not traditional EE targets. Energy efficiency policy was aligned with the thinking of Amory Lovins, who argued that changing the hardware to provide the same energy services, but using less energy (cold beer and warm showers were his examples), would suffice to bring about the needed demand reduction. The question of what people were using energy *for* was quite irrelevant. The L-word (*lifestyle*) was rarely used in EE policy discussions, as it invoked the by-then notorious image of President Jimmy Carter wearing a sweater in the White House as he asked for personal sacrifice in dealing with an energy crisis (he characterized it as the moral equivalent of war). Carter's call for Americans to *sacrifice* has been blamed for the subsequent defeat of his re-election campaign.

The EE movement maintained a great distance from discussions of lifestyle and behaviour change throughout the 1980s, but the advent of market transformation in the 1990s actually did open rhetorical space for these topics. After all, a thorough understanding of any market for energy-using goods would require some understanding of the uses to which the goods and energy were being put – as well as an understanding of developmental trajectories and alternatives. At the very least, some insights from marketing would be useful in differentiating consumer preferences across subgroups. However, by 2001, little of this sort of work had been done in the market transformation context. Therefore, the EE programs could only increase support for their conventional offerings, and the Flex Your Power campaign had to rely on creativity and luck as it navigated novel terrain. Trying to tap into consumer concern, it risked generating more apprehension, anger, and possible backlash; it was a gamble, but one that could not be avoided.

The conventional wisdom in EE circles was that consumers would not respond positively to threats to their quality of life and that self-indulgent Californians in particular might be liable to revolt.[11] Myths and imaginings aside, electricity demand in California peaks in the heat of summer, and asking persons to make do with less cooling for the common good was not something that anyone wanted to do – and, in fact, no one did ask for that; officials only suggested that setting thermostats up a few degrees might be something to consider. No one had any idea what the consumer reaction might be. The smart money was on disgust and possible disorder.

What Was the Response from Consumers?

In late 2000, I was asked by the California Energy Commission to evaluate consumer response to the coming policy changes, including ramped-up EE programs, appeals to the public, and advertising. The primary research questions were who would do what, and why. The research team conducted in-depth interviews with large energy users (government agencies, firms, and agriculturalists), as well as media inventories, two waves of residential consumer surveys (both during and one year after the crisis), and analyses of consumption data, weather, and price effects. The many sources of simultaneous influence on consumers made it impossible to attribute cause precisely. However, we were able to detect changes in consumption at the household level, rule out weather as the cause (it was actually warmer during 2001 than the previous year), and find significant behaviour change in the absence of price changes. The analysis and findings are detailed in Lutzenhiser et al. (2003).

The results were far from what would be expected from a 'selfish, rational consumer,' the model on which traditional EE policy was based. We found significant reductions in energy use, which were motivated by a combination of civic, economic, moral, and altruistic concerns. The conservation behaviours were quite varied – and variable across the population – and a number were creative contributions that had not been cued by advertising. There was a surprising incidence of extreme conservation behaviours (including shutting off air conditioning in very hot conditions), persistence of some behaviours a year after the crisis, and new perceptions and beliefs about the nature of the electricity grid, future problems, willingness to act, and lifestyle change.

A large number of households were interviewed in late summer and early fall of 2001. Respondents who indicated that their energy-using practices had changed in any way as a result of the recent energy situation were asked to describe those changes in their own words. The responses were classified into nearly one hundred different types of conservation behaviours. A compressed set of categories is shown in table 12.1.

Conservation actions were widespread across households. More than 75% reported taking at least one conservation action, and more than half (58%) took two or more such actions (the average was 2.4). Figure 12.2 shows the distribution of types of conservation behaviour reported. It also includes responses from a second survey, taken a year later (discussed below). Using less lighting was the most common response. In all, about 10% reported using no air conditioning at all, and 48.5% took other conservation actions related to cooling or heating. Other actions reported by 20%–30% of households included small-equipment behaviours such as turning off equipment when it was not in use, using compact fluorescent or other lower-energy bulbs, and shifting energy use to off-peak hours.

Although hardware changes were heavily promoted in public appeals, relatively small numbers reported making major energy efficiency investments in their homes such as shell improvements or replacing inefficient appliances. This was especially true for low-income households, apartments, and other rental units, where such actions were

Table 12.1. Reported Conservation Behaviours

Shell improvement	Hardware-related one-time improvements to the house (e.g., windows, insulation, a new piece of fixed equipment such as a water heater, AC, furnace)
Light bulbs	Hardware-related purchase/use of compact fluorescent bulbs or other energy-saving or low-wattage bulbs
Appliances	Hardware-related purchase/use of new non-fixed appliances (e.g., refrigerator, washer-dryer, window AC, fans)
Lights behaviours	Behaviours related to turning off lights or using fewer lights
Small-equipment behaviours	Behaviours related to household appliances (e.g., turn off, use less, unplug)
Large-equipment behaviours	Behaviours related to pools, spas, irrigation motors (e.g., turn off, use less often)
Not-using-AC behaviour	Behaviour related to not using the AC at all
Other heat/cool behaviours	Behaviours related to heating and cooling other than not using the AC at all (e.g., use AC less, use ceiling fans, draw curtains, vent at night, turn thermostat up or down)
H_2O behaviours	Behaviours related to using less water or using less hot water (e.g., shorter showers, wash in cold/warm water, turn water heater down)
Peak behaviours	Behaviours related to using energy during off-peak hours (e.g., for washing, cooking, cleaning)
Vague behaviours	Behaviours that were stated in general terms (e.g., be an over-all conserver,' 'be less comfortable,' 'use little energy')

virtually non-existent. Behavioural changes predominated. Among hardware purchases and investments, the installation of compact fluorescent lamps (CFLs) and other low-energy bulbs was the most common, reflecting the fact that it is the easiest hardware action for households to take.

As mentioned, the most prevalent cooling conservation message during the summer of 2001 was to set air conditioning (AC) thermostats up (to 78 degrees or higher), rather than to not use AC at all. The results of the survey show that among households with central AC, 36% reported using less or no AC, while 29% of those with room AC reported using it less or not at all. What was surprising was the propensity among those consumers to turn off AC completely, as only a small fraction of households went for the easier turn-up option. The actual verbatim re-

Figure 12.2. Proportions of California Households Reporting Various Conservation Behaviours in 2001 and 2002

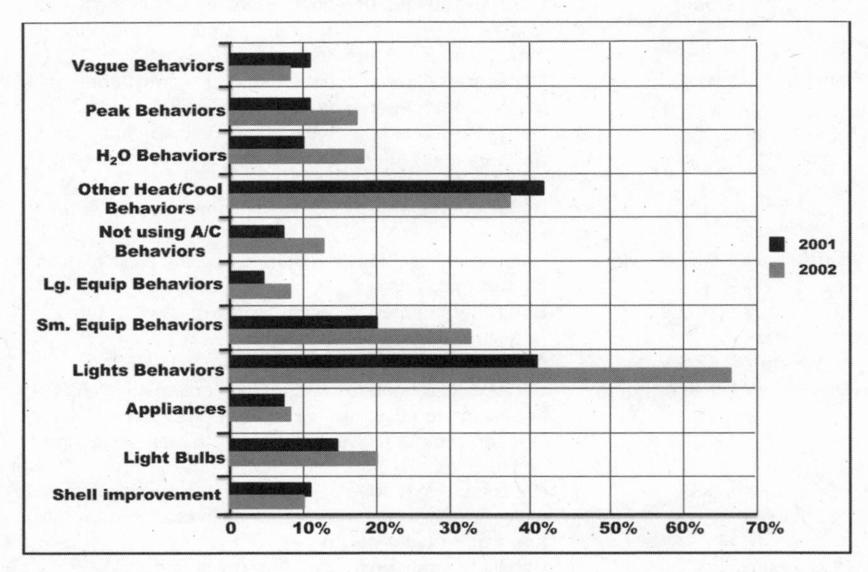

sponses in this category describe actions such as 'draw window shades or curtains during the day,' 'turn thermostat off when I'm away,' 'don't use the AC,' 'use the air conditioner less often,' 'use it only when we have to,' 'open windows at night,' 'open windows in early morning,' and 'close off part of home to use less cooling.' These and other cooling conservation behaviours likely delivered great energy and peak demand benefits, since cooling accounts for 35.5% of peak megawatts and 7.4% of annual residential consumption (California Energy Commission 2003b).

In terms of gross energy savings, Goldman, Eto, and Barbose (2002) estimated that energy efficiency and onsite generation projects initiated in 2001 would account for about 1,100 megawatts of customer load reductions once all projects were installed. These savings represent about 25% to 30% of the observed load reductions, with the balance attributable to conservation behaviour. Our own analysis of household level consumption found that energy savings were somewhat concentrated, with a subset of the population making larger reductions. Also, some utilities were more likely than others to have concentrations of these 'super savers.'

Motivations and Lifestyles

Consumers were certainly interested in containing costs. However, their reported motivations for conservation do not bear out the conventional wisdom in energy policy circles that conservation action can only occur with price increases.[12] During the 2001 energy crisis, actual price increases were sporadic and unevenly applied and often came long after the conservation action had been initiated. Instead, it seems that for most people the motive to conserve was a combination of monetary and non-monetary concerns (see figure 12.3). For some, behavioural change had nothing at all to do with an interest in saving money or the anticipation of price increases but was motivated by civic concerns and altruistic motives. While most conservation actions could produce cost savings (and, in several cases, additional rewards from the utility if the savings levels were high enough), about 20% of households reported shifting energy use to off-peak periods, even though it meant no net savings on their bills.

Many households said that they planned to continue their energy conservation actions. This is not surprising, since almost 60% reported that those actions had no serious effect on their quality of life, and 18% even experienced an improved quality of life. In the second survey, conducted in 2002, a majority of households reported actually continuing a variety of conservation actions. We found that 90% of the households that had reported taking conservation actions in the summer of 2001 were still pursuing at least one action a year later.

Voluntary conservation actually continued to produce energy savings in 2002 (see figure 12.2). Analysis of system-level data, adjusted for differences in weather and economic activity, showed reductions in electricity demand in 2002 relative to 2000 that were about half as large as the 2000–2001 reductions (California Energy Commission 2003a).

Longer-Term Considerations and Effects of the Crisis

Consumers reported a continuing concern for the California energy situation, a willingness to continue conserving energy, and a seriousness about their commitments. In both the 2001 and 2002 surveys we asked, 'Which statement comes closer to your view: "Californians can maintain their lifestyle, and the state's energy problems can still be solved," OR "Californians must make real changes in their lifestyle in order for the state's energy problems to be solved"?' The somewhat surprising

Figure 12.3. Motivations for Conservation Action

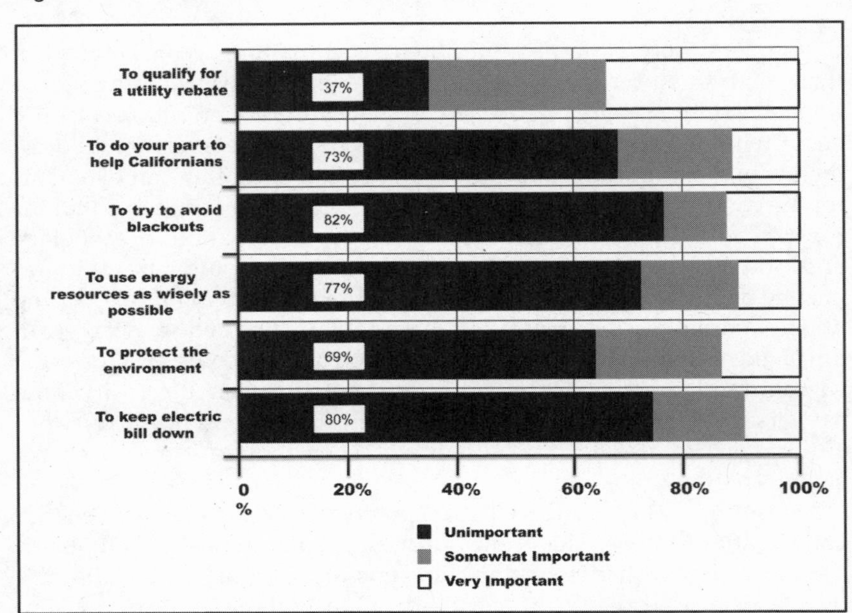

results were that large majorities (66% in 2001, and 71% a year later) believed that lifestyle changes would be required in order for energy problems to be effectively addressed. This response does not necessarily reflect a personal commitment to lifestyle change; rather, it is one of general view – a respondent may think that some lifestyles have to change but not necessarily his or her own. However, we also have evidence from questions that speak more directly to personal values.

Consumer attitudes about energy conservation were explored in greater depth in the 2002 survey, which found that Californians continued to believe that energy problems were serious and that conservation was still important. As shown in table 12.2, very large majorities (80%–93%) offered strongly pro-conservation views.

The Energy System 'Problematized': Emerging Consumer Views of Energy Issues and Energy Policies

Myths rooted in the 1970s zeitgeist encouraged fear of consumer backlash and kept voluntary conservation and lifestyle change off the table

Table 12.2. Energy Attitudes

Personal conservation attitude statements	Agree	Disagree
'I really don't care much about energy and see little reason to conserve.'	8%	92%
'Even if I cared about energy, there is not very much any individual can do to conserve that will have much effect in the long run.'	20%	80%
'We could all use a lot less energy than we do, and if many people conserved, we could all make a big difference overall.'	88%	11%
'Regardless of whether it makes a difference, everyone has a moral obligation to do the best they can to conserve energy.'	88%	11%
'It makes sense every once in a while to ask citizens to reduce their energy use in order to do their part to avoid blackouts and keep costs down.'	93%	7%
'It is worth it to pay more for energy in order to never be asked to conserve.'	12%	88%
'My conservation efforts over the last few years have involved real sacrifices.'	40%	59%
'As energy prices increase, everyone will become a conserver.'	52%	47%

in EE policy discussion. The unintended experiment of the California 2000–2001 electricity crisis proved those notions wrong. While it is risky to generalize from the events surrounding a single crisis in a single state, we do now have a much better sense of what consumers might be willing and able to do under some unusual circumstances. Moreover, while it may be true that Americans do not want to hear that the energy system (and the lifestyles that it supports) are vulnerable, in California at least the modern electricity grid has now been 'problematized.' For Californians, it has taken its place in the pantheon of problems of modernity, next to a clogged and dangerous highway system, air pollution, tenuous food supply chains, the rapid spread of exotic diseases, environmental degradation and ecosystem decline, crime and crowding, and so on – all grand systems failures that are seemingly intractable.

Figure 12.4 shows the results of a series of survey questions about the concern for energy-system-related problems. A clear majority felt that all these problems were serious and would continue to be serious in the future. They included shortages of energy imports, transmission

Figure 12.4. Views of the Seriousness of Future Energy Problems

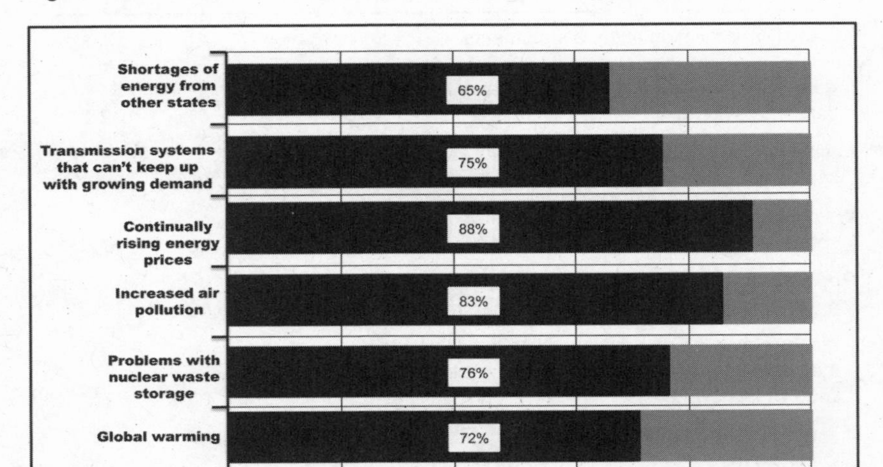

system limitations, continually rising costs of energy, increased pollution, nuclear waste storage, and global warming.

Rather than resignation to the situation, though, consumer responses to another series of questions about energy policy options indicated strong support for proactive institutional efforts to address these problems. By wide margins, consumers said that it was 'very important' for the state to continue to support energy conservation programs (59%), encourage energy efficiency (75%), and expand the use of renewable energy (73%).

Summing Up the California Experience

It is unlikely that anyone in California could have predicted the significant system-level reductions in electricity use that resulted from the conservation actions of millions of residential consumer households in 2001. After all, conventional wisdom in the energy community has been that households are relatively fixed in their demands. Likewise, conservation behaviours have been viewed as unreliable sources of energy savings and peak load reduction. Yet, the experience from 2001 indicates that electricity demand reductions from conservation behaviour can be significant.

This revelation suggests a need to develop a better understanding of household energy behaviour so that it can be reflected in future energy policy and program efforts (Kunkle et al. 2004). The traditional EE policy focus on hardware efficiencies is clearly too limited – and even unhelpful – when we see devices in context and recognize that variability in use translates to variability in resultant energy demand. Small changes in behaviour can have, in the aggregate at least, demonstrable effects. We have also learned that energy conservation behaviours are not rare and set apart from otherwise routinely profligate behaviour. Conservation in households is fairly widespread and continually evolving. Although we still have a limited understanding of the organization and dynamics of everyday energy-using action, we can see now that energy conservation behaviours are part of how households routinely manage energy use. Finally, we now have some evidence that the ability of households to act can be enhanced by external influences, although we know little about the nature of effective influences or the capacities and conditions that limit them.

New Policy Potentials and Participants

Taken together, the lessons from the California crisis experience are remarkable. The consumer response was broadly based, sober, and creative, and the effects were significant in terms of system stabilization. Policy myths were called into question. Public views of the energy system were altered. And subsequent developments have shown that a new space has been opened in policy discussions to take consumer behaviour seriously.[13]

In the remainder of this chapter, I consider the problems inherent in applying understandings of human action that have long informed device-centred EE policy to the problem of climate change. I also look at the potential of the social sciences and related energy disciplines to provide a more accurate and actionable perspective on human behaviour and choice. Finally, I assess the policy challenges ahead in terms of both promising new developments and serious barriers to policy innovation and effectiveness.

Applications to Climate Change Mitigation

The accomplishments of several decades' worth of west coast EE policies should be celebrated. Advanced building codes, appliance standards, and concerted programmatic efforts were put in place. Through a combination of these measures, per capita electricity consumption in California has remained essentially flat over the past thirty years while U.S. per capita consumption has grown substantially.[14] Energy efficiency has been institutionalized as a low-cost and reliable source of energy supply in the region. Advanced program evaluation expertise has evolved to estimate savings, part of an energy efficiency industry that has grown to manage the very large sums of money now being invested in EE along the west coast.

Energy efficiency will continue to be an important source of greenhouse gas reductions, as significant additional EE gains are expected. However, the new energy-policy challenges that we face are larger than EE advocates had ever imagined (or than EE was ever intended to address) and will require grander solutions. Climate change mitigation calls for significant changes in how we convert and use energy, and EE is only part of the story – one wedge in Pacala and Socolow's (2004) analysis. California, Oregon, and Washington have set very ambitious carbon emissions reduction goals: as high as 80% below the 1990 levels by 2050. The best available household consumption data from the California Energy Commission's demand forecasting model – using the most optimistic scenarios – raise serious questions about our ability to come close to those goals in the residential sector with our present policy arsenal (Rufo and North 2007).

Much more is needed, and the scale of the problem is daunting. Dramatic improvements in new buildings (well beyond current codes and practice) are required. Maximum retrofit of existing buildings is necessary, using the most advanced technologies available. Vast improvements in end-use technologies are called for to replace hundreds of millions of devices on the west coast alone. Energy efficiency policy was never intended to undertake the mass transformation of complex systems of buildings, equipment, vehicles, behaviours, settlement forms, infrastructures, and lifestyles. People and institutions need to be able to organize to take on the challenge of a massive overhaul of hardware. They also must be ready to change their habits, expectations, needs, and lifestyles – and the California crisis experience gives us reason to be hopeful that they might be.

There is cause for concern that conventional device-centred EE thinking and policy and program offerings are as ill suited to the present carbon crisis as they were to the California electricity crisis. We need new ideas, but at present our toolkit is sparse. There is renewed hope for markets – namely, the power of carbon markets to unleash restrained ingenuity. We know that markets do that, but we also learned during the deregulation era that we cannot always control the direction of the ingenuity unleashed. We know the power of EE programs, but we also understand the limitations of their incrementalism in the face of problems dire and immediate. The narrow focus of EE policy perspectives on hardware and their corresponding ignorance about people – alternately invisible, 'average,' and interchangeable or robotically rational and calculative – has been discredited. As we have ample evidence of highly variable behaviour and limited economic influence, we know the facts to be different.

We have the advantage now of learning from the California experience in order to build upon some new insights about people, energy, and systems. The problem is that our knowledge of routine everyday behaviour, conservation action, the nuances of technology choice, the policy process, large-scale intervention, technological innovation, supply chain dynamics, and a host of related issues in the human-energy-technology-environment system is still quite limited. The system as a whole – sometimes called a 'coupled natural and human' system (National Science Foundation 2007) – has only recently become the object of investigation by atmospheric scientists, biologists, and economists.[15] The focus is still at highly aggregated and macro-geographic scales, where current understandings of behaviour are sketchy at best and more often nonexistent.

Conceptually, the household consumption system consists of a fairly large number of dissimilar, but interrelated and interdependent, elements that operate simultaneously on multiple levels. The mixed social, technological, and environmental elements in this system have long been recognized in energy analysis and environmental social science.[16] Several theoretical models have been constructed over the past twenty years, and a recent review by Keirstead (2006) considers their past applications and current status. One of the best in-depth theoretical treatments of the social organization and dynamics of these systems can be found in Shove et al. (1998) and Wilhite et al. (2001). Figure 12.5 is an oversimplified attempt to capture the interacting elements of natural environment, weather, building, equipment, human occupants,

Figure 12.5. The Household Consumption System in Context

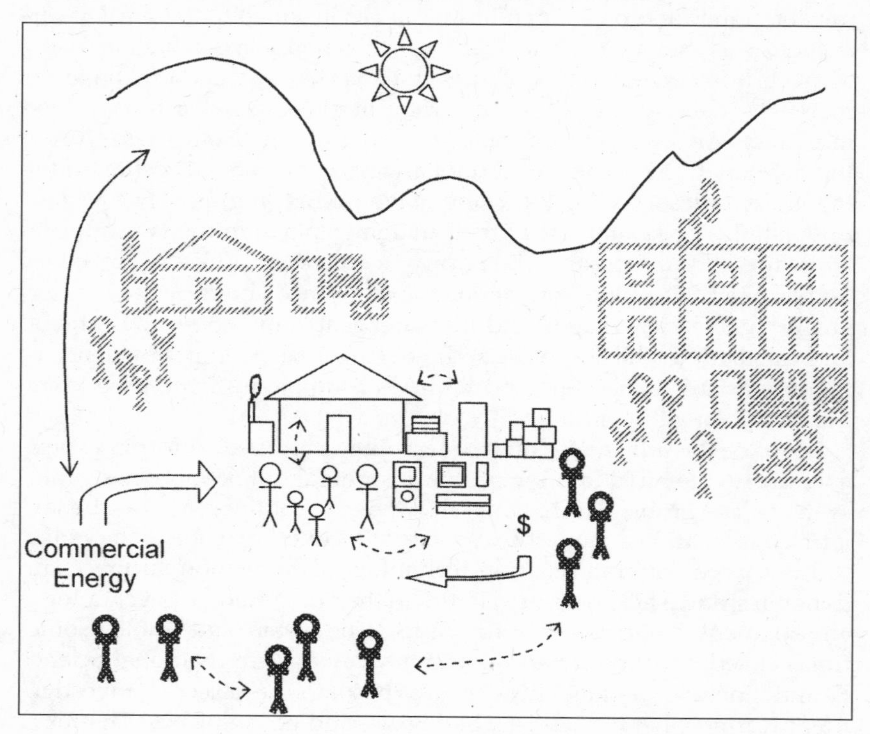

surrounding built and social/cultural environments, institutional influences, energy provision systems, and supply chains. It is offered as a heuristic to keep in mind as we consider how to improve our understanding of the system.

Potential Contributions from the Social Sciences and Other Academic Disciplines

The social sciences should be expected to provide insights that can help us better understand, and intervene more effectively, in complex systems such as these. In fact, considerable relevant knowledge has been accumulated in the academic social sciences and in applied energy investigations carried out by university-based engineers and economists. There are alternatives to the device-centred view of people and technologies, and there are perspectives that stress complexity and dynamism, in contrast to the mechanical views of the physical-technical-economic model. This is on top of the wealth of knowledge proffered by EE program operators and evaluators, which are accessible through several applied EE interest organizations that maintain publishing outlets for this information.[17]

While much of this applied work is quite detached from cutting-edge scientific developments, academic researchers are, in turn, distanced from the world of EE applications, and increasingly so. When the American Council for an Energy-Efficient Economy (ACEEE), an applied interest organization, was organized in the early 1980s, most of its first members were national laboratory research scientists, university-affiliated faculty, and students. It was part advocacy group, part information exchange, and part social movement. As EE became institutionalized and the EE industry emerged, academic interest waned. A generation of researchers left the field (particularly in the social sciences), and dwindling numbers of new students were trained to replace them. In truth, except under large-scale crisis conditions, energy is of very minor interest to the mainstream social sciences, considered either 'too applied' (versus 'basic theoretical') or simply low on the list of disciplinary concerns. Therefore, it should not be surprising that by 2008 only twenty to thirty of the more than nine hundred ACEEE conference attendees were even affiliated with universities (and most were engineering-oriented researchers).

The disciplines that need to be engaged in a new phase of EE thinking related to complex systems and climate change mitigation include (1) the standbys, engineering and economics; (2) the social sciences, including sociology, anthropology, and psychology; (3) the professions that draw upon all of these (that is, public policy, business, education, and technology management); and (4) interdisciplinary fields, such as ecology, geography, environmental studies, urban studies, science and technology studies, and sustainable development. All have a good deal to contribute. By way of example, table 12.3 summarizes some of the relevant areas of study that can readily be applied to the problem of EE in complex systems. There are, in fact, useful small literatures in each of these areas. However, the work of integrating them remains.

This is not a straightforward task, nor one that aligns nicely with academic and disciplinary priorities and reward systems. If energy and behaviour as an area of scientific inquiry is the apple in figure 12.6, then each of the disciplines takes a bite from it. Much of the apple belongs to

Table 12.3. Disciplines and Conceptual Interests Relevant to Energy Efficiency

Sociology	Households, organizations, communities, institutions, status, technological change
Anthropology	Cultures, folk models, meanings, practices, habitation
Psychology	Attitudes, motivation, information, perception, behaviour change
Economics	Pricing, financial incentives, benefits
Engineering	Hardware controls / human interfaces, system operations

Figure 12.6. Disciplines, Relationships, and Bites of the Apple

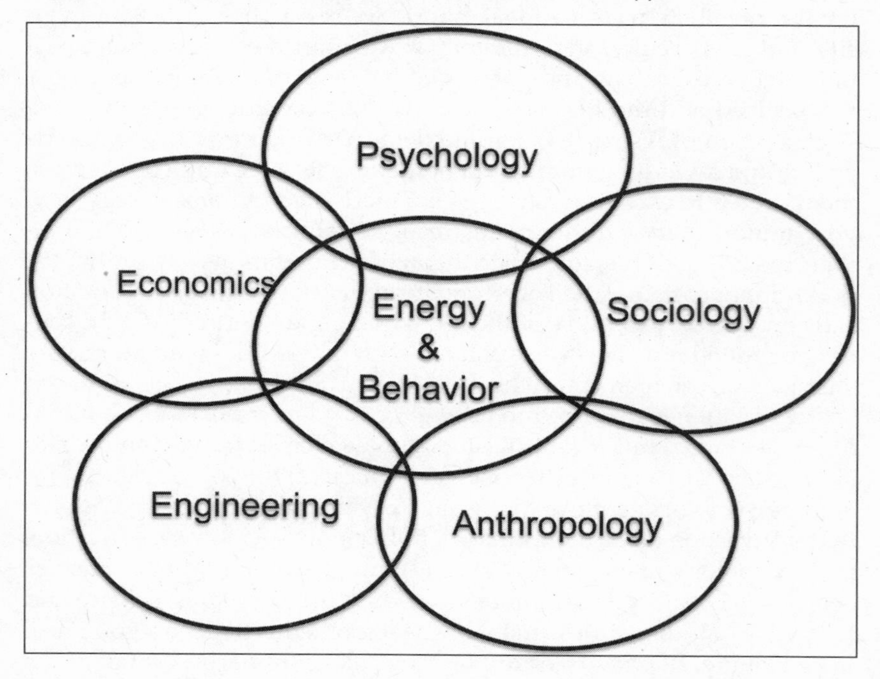

no one. Each discipline has some shared interests and connections with disciplines that are adjacent in the graphic, but is quite disconnected from others. There are very real differences among disciplines in methods and theory. There are conflicts of perspective and interpretation, but different disciplines may actually be saying similar things in different ways. There may be complementary insights across disciplinary divides, as well as gaps in knowledge that only collaboration can fill. Disciplinary organization does not make this easy, but it needs to happen – and quickly – if the social sciences are to contribute insights to the mammoth policy dilemma of how to drastically reduce greenhouse gas emissions.

Promising Policy Developments and Remaining Problems

On the policy side, we now see some promising EE innovations that merit consideration. Some of these are centred on the U.S. west coast, but many now are national in scope. Developments are also occurring rapidly in Canada, Europe, and beyond. In fact, the center of EE policy innovation may be up for grabs as European initiatives take hold. However, innovations are alike in that they are hindered by persistent gaps in our shared knowledge.

The first important innovation is *demand response* (DR). The problem of peak demand has become a permanent fact of life in California, where periodic system operator alerts are issued when projected capacity is very close to expected demand. Whole customer classes are being equipped with interval electric meters that can be read remotely and will collect information on energy consumed during high-demand time blocks, such as late afternoons in the summer. Higher prices can be charged at those times. Also, during times of critical peak where alerts might be issued, much higher prices (as much five to ten times the standard rate) can be charged to discourage energy use and compensate for the much higher costs of electricity on short-term markets. In order for these innovations to be successful, a much better understanding of consumer behaviour, knowledge, constraints, and support requirements is needed. Efforts are underway by California regulators to assure that understanding is improved before large-scale adoption of demand-response programs and rates.

A return to *market transformation* is taking place in California. Taking MT seriously will require a much better understanding of the social actors and systems involved in the energy use and choice in these settings than conventional EE device-centred and economic-rational frameworks require.

The University of California Institute for Energy and the Environment is pursuing some *disruptive technologies* through the R&D activities of California public-purposes technology. These are technologies that could yield ten times the energy savings of current approaches, at one-tenth the cost. A number of university researchers and corporate partners are involved.

Zero net energy buildings are the goal of several initiatives sponsored by the U.S. Department of Energy and selected for policy support by the California Public Utilities Commission (CPUC). The former is supporting R&D in the area. The latter has mandated that a combination of high-efficiency design and equipment, along with solar, wind, or other renewable energy, be used so that all new buildings are zero net energy – by 2020 for all new residential buildings and by 2030 for all new commercial buildings.

The CPUC has recently commissioned a series of studies and white papers *exploring the behavioural assumptions* that underlie current EE potential analyses, evaluation approaches, program designs, and customer segmentation strategies. The goal of this work is to bring to light the assumptions, such as economic rationality, knowledge limits, and stylized decision processes, in order to determine whether improved understandings of consumer behaviour and choice can be incorporated into EE policy development and program planning.

Discussions have been conducted among utilities and regulators in California concerning whether *voluntary behavioural conservation* (of the sort evidenced during the California crisis) might be considered as a significant resource on par with hardware-focused EE programs and new power acquisition. Issues involve appropriate approach, consumer autonomy and rights, measurement, persistence, and valuation.

As lifestyle has become a more acceptable topic in EE circles, the dilemma of *sufficiency versus efficiency* has been raised at recent EE conferences (for example, Rudin 2000, Harris et al. 2006, Princen 2006). These discussions have asked how much is enough; that is, what are reasonable, adequate, or sufficient levels of consumption?

Analysis of *interactions among end-use sectors* involves serious consideration of trade-offs in energy use between home, work, transport, and public places. A highly efficient dwelling can become the locus of increased work activity, which may increase its energy use, although energy use at the workplace may be reduced. Mass provisioning at restaurants may offer cooking, cleaning, and heating/cooling efficiencies on site and at home, but greater energy use would be required for travel. Energy planning and programs are conventionally sectoral (residential, commercial, et cetera), so considering multi-sectoral interaction is a novel pathway to improved *net-efficiency gains.*

Finally, some advanced planning efforts consider the possibilities for energy reductions resulting from long-term changes in *settlement patterns, lifestyles, and built environment*. This involves systematically examining questions related to end-use sector interactions as well as fundamental questions about significantly different ways to live in the future. Several formal models have been developed in this area (for example, California Energy Commission 2008; Metroquest 2008). British Columbia Hydroelectric included this sort of policy modelling in its most recent assessment of energy efficiency potentials for the province.

Impediments to Policy Change and the Way Forward

All of the policy developments mentioned incorporate a systems approach to thinking that takes behaviour as seriously as hardware, buildings, and the environment. Taken together, they make possible a cautious optimism for the prospect of significant change in energy use patterns and overall efficiencies. They may even suggest a future for large-scale carbon reduction initiatives. At the same time, we face a number of significant challenges to that progress.

The inertias of the systems of consumption and provision cannot be underestimated. It will be important to understand how change in complex socio-technical systems takes place. Interdependencies, ideas, paradigms (like the conventional PTEM mentioned at the beginning of the chapter), institutions, regulatory structures, pre-existing policies, and interests are all at play – and all have potential to slow or even block changes. Technologies themselves are interlinked and must be constantly manned in order to operate. This demands vast resources, time, and attention – all of which are in short supply but are needed to bring about system alteration.[18] Our analytic tools and techniques can allow consideration of alternative futures, but they also can produce narrowed views and 'blind spots in policy analysis' (Stern1986).

The very organizational matrix through which we will launch progressive energy activities can inhibit success because of commitments to antiquated ideas, approaches, devices, and delivery systems that were developed under very different conditions. By institutionalizing and industrializing EE, we have achieved some important successes in slowing the growth of consumption, but they come perhaps at the cost of our ability to move rapidly and effectively in a broadened sphere of activity that now involves consumers, systems, and the global climate.

There will be efforts to simply extend conventional EE thinking to human consumers and systems by adding amendments to the PTEM.

An emerging case may be the interest in marketing to individuals, particularly social marketing that attempts to change individual behaviour through message manipulation and by couriering information through social networks. Some of this may be successful, but it is also possible that the underlying perspective that focuses on individuals is fundamentally wrong. It reduces the complexity of action and choice by multi-actor households embedded in dense socio-technical contexts (in which institutions and supply chains are also very powerful actors) to individual choice. Even if that choice is seen as involving social norms and influential individuals in networks, the framework may distort more than it explains.

Large-scale change will be much more difficult to accomplish. We really do not know how to do it. We know that the systems are complex, the problems are complex, and elegant solutions will be required. This need not mean intimate involvement in lives and groups. It is possible that some relatively straightforward actions can help to motivate change and amplify it in positive ways throughout systems. For example, getting the prices right by incorporating externalities can help. But, again, we have seen that market creativity is not always the creativity that we need. Therefore, solutions will need to thoughtfully link markets with regulation, intervention, and voluntary action, borrowing insights from across the academic disciplines. Behavioural changes will be required from persons, households, communities, and organizations. Technologies both new and rediscovered must be harnessed, along with their accompanying practices, habits, and norms (Shove 2004). The combined contributions of government, business, university-based science, the non-profit sector, and citizens will be required. We have no idea whether this is possible. There are some hints from the successes of past collective actions taken in the face of threats (for example, during war time) that much can be achieved, but the scale of our task – whether it is mitigation or only slowing climate impacts enough to help make large-scale adaptations possible – is daunting. However, we have no choice but to try.

Conclusions

The aim of this chapter has been to review developments and issues related to energy efficiency and consumer behaviour and to identify some insights into both from consumer responses to the 2001 California electricity supply crisis. No definitive conclusions are offered, but enough is known to reasonably ask how applicable those experiences are to the larger issues of climate change, and how the model might be translated to suit the conditions of other places and policy contexts.

While detailed answers to these questions are also premature, the events and policy developments reported here are highly suggestive. First, they reinforce our understanding that policies do not appear de novo but evolve in specific historical and institutional contexts. The rise of energy efficiency policy in the regulated electricity sector on the west coast of the continent clearly illustrates this. As a result of local and regional environmental concerns, conservation was reframed as efficiency, and demand side management was conceived as a new source of energy supply. Over time, a significant efficiency industry developed, and policies settled into a device-oriented perspective on consumption and efficiency-improvement that addressed regulatory needs. These policy arrangements and efficiency supply systems have grown in sophistication and now also serve as a source of ideas and expertise to other policy contexts in North America and beyond.

There are important lessons in this for would-be translators of the west coast experience. For example, the system that ultimately appeared was not one that the original conservation advocates had imagined. It has a narrow focus and a track record of unquestioned success, but also one of modest incremental efficiency improvements. The legal requirements and institutional arrangements around electric utility regulation have produced a formalized model of consumer behaviour that seems fundamentally at odds with reality. Yet the scientific and technical expertise that should serve as a corrective is complicit (particularly in the cases of engineering and economics), and the social sciences have offered little in the way of useful correction. It is difficult to imagine this particular system scaling up to deliver large quantities of efficiency improvement on anything like the temporal, spatial, and organizational scales required to seriously address accelerating rates of climate change. Some of the new market transformation initiatives may offer improved prospects for rapid improvements in efficiency. But the problems of policy design and program delivery are formidable.

Second, the significant response of consumers to the California energy crisis holds the seeds of hope in several senses. The conventional understandings of consumers were found not only to be inadequate and incorrect, but misleading as well. Devices were seldom replaced as an energy-saving strategy. Behaviour was key. Altruism appeared to be as important as self-interest. Conservation was a virtue. Lifestyles were not non-negotiable or a politically toxic topic of conversation. Energy-savings action was taken, was effective, was sometimes bold, and persisted for a time following the crisis. Energy and the supply system were brought into the realm of public problems – eroding at least a bit their invisibility and mysterious workings. California's long-term energy problems were not solved. However, the crisis enables a possible reframing of consumers as active managers of their own energy use and members of a social system. It also opens policy doors to new ways of thinking about costs, benefits, risks, shared responsibility, communications, incentives, and particularly the role of behaviour in energy use – and of behaviour change in efficiency improvement.

If market transformation approaches are developed that take seriously the need to simultaneously reconfigure technologies, buildings, supply chains, regulatory environments, commercial priorities, and consumer behaviours and demands, then a new, richer understanding of consumer actions and roles will also be necessary. Therefore, from the experience of system distress and crisis we may have found the beginnings of new ways to begin to re-imagine consumers and consumption so that energy policies can be more nuanced and effective.

NOTES

1 In these amendments, individual 'attitudes' and 'influences' are imagined to cause behaviour change. The empirical performance of individualistic behaviour models has most often been poor in the energy case, and a compelling argument can be made that the phenomena we are most interested in changing (that is, energy-use behaviour and technology-related choices) are not organized at the levels of individuals or networks of individuals.

2 See Lutzenhiser (1993), Wilhite and Lutzenhiser (1999), Lutzenhiser (2002), Biggart and Lutzenhiser (2007), and Wilson and Dowlatabadi (2007) for at least partial explanations.

3 It should be noted that the federal *National Energy Act* (NEA) of 1978 provided reinforcement. The NEA required states to offer some level of residential energy conservation support, and it also included provisions to encourage small independent power production. Ironically, the latter unintentionally created an opening for large non-utility electricity-generating firms to later demand entry into utility markets (Hirsh 1999).

4 The technology is essentially a stationary natural-gas-fuelled jet engine connected to a generator.

5 The terms *deregulation* and *restructuring* have been used interchangeably in the discussions about opening regulated electric utility businesses to greater market exposure. In this chapter, I use the term *deregulation* because I think that it may more accurately capture the historic aims of the movement. However, I do not intend to imply that the goal has ever been to remove all regulation, but rather to reduce levels of oversight and regulation of electric utilities. Just how this has been done across the United States is enormously varied (Energy Information Administration 2003). However, through this process, in many states, utilities now have somewhat greater exposure to market risks and benefits than they did in the early 1990s.

6 A range of market transformation efforts were launched on the west coast and elsewhere, although they were often simply repackaging and renaming bits of earlier EE policy and practice. One of the seminal published statements of market transformation principles was Geller and Nadel (1994). In the Northwest, a new MT-focused organization that was governed and funded by four member states, utilities, and the federal Bonneville Power Administration was created. This Northwest Energy Efficiency Alliance (NEEA) continues as a collective effort to intervene strategically in markets for energy-using equipment, buildings, and infrastructure. For a statement of how market transformation ought to be planned, delivered, and supported by key research efforts, see Blumstein, Goldstone, and Lutzenhiser (2000). Archives of MT-focused work can be found at the websites of NEEA (http://nwalliance.org) and the California Measurement Advisory Council (http://calmac.org).

7 This investment recovery arrangement, coupled with partial reliance on spot markets to provide crucial increments of power, is what led to the electricity supply crises of 2000 and 2001. For a detailed explanation of the roots of the crisis, see Borenstein (2002).

8 The regulated utilities prefer to call this public-purposes surcharge (approximately 3%) *a tax* (that is, on their utility bills) but insist that since they collect it, they are the most appropriate agents for spending it.

9 The best explanation for the abandonment of MT after the crisis may be a newfound distrust of markets in any form.

10 The state of Oregon, in particular, offers a fairly wide array of EE (and renewable energy) services, incentives, and tax credits through several public and quasi-public agencies; it also contributes to the regional efforts of NEEA.

11 In fact, when energy bills increased suddenly in San Diego in the summer of 2000 to reflect rising wholesale electricity prices, the political reaction was so intense that the regulators stepped in and capped retail prices –

hardly what might have been expected to occur in competitive markets, where prices were supposed to fall.

12 A recent review of the economics literature on the price elasticity of household energy demand shows quite ambiguous results, with confidence only that increasing prices are likely to be accompanied by conservation action and efficiency choice in the long run (Kriström 2006).

13 For example, the recent 'behavioural summit' of government officials, utility executives, and academics (California Institute for Energy and Environment 2007); the Behavior, Energy, and Climate Change conference held in Sacramento in November 2007 (American Council for an Energy-Efficient Economy 2007), which is planned to be repeated annually; and the commissioning of a series of white papers on behavioural topics by the California Public Utilities Commission (CIEE 2008).

14 See Bernstein et al. (2000) and Kandel, Sheridan, and McAuliffe (2008) for a discussion of this effect.

15 For example in the Intergovernmental Panel on Climate Changes' assessments (IPCC 2008).

16 For example, see Cramer et al. (1985), Lutzenhiser and Hackett (1993), and Lutzenhiser and Bender (2008).

17 The most important are the annual proceedings of the International Energy Program Evaluation Conference, and the biennial proceedings of the American Council for an Energy-Efficient Economy and the European Council for an Energy Efficient Economy conferences. However, it is important to understand that the work reported in these proceedings, while peer reviewed (mostly by fellow program managers and evaluators), tends to be largely descriptive and often represented as work in progress. This 'grey literature' is still valuable, but primarily as a source of data, since it is generally not intended to contribute to the development of theory and, in fact, is theoretically derivative from the academic sciences.

18 The literatures on innovation and change in complex technologies, including human behavioural roles in design, operation, and alteration of socio-technical systems, will be quite useful in serious attempts to expand EE and other efforts to reduce carbon emissions. Some of this work has explicitly considered energy and consumption technologies, for example, Hughes (1989), Bijker and Law (1992), Shove et al. (1998).

REFERENCES

American Council for an Energy-Efficient Economy (ACEEE). 2007. Behavior, Energy and Climate Change Conference. Sacramento, CA, 7–9 November.

http://www.aceee.org/conf/07becc/07beccindex.htm (accessed 9 September 2008).

Bernstein, M., R. Lempert, D. Loughran, and D.S. Ortiz. 2000. *The public benefit of California's investments in energy efficiency*. RAND Monograph Report MR-1212.0-CEC. Santa Monica, CA: RAND.

Biggart, N., and L. Lutzenhiser. 2007. 'Economic sociology and the social problem of energy inefficiency.' *American Behavioral Scientist* 50:1070–87.

Bijker, W., and J. Law. 1992. *Shaping technology / building society*. Cambridge, MA: MIT Press.

Blumstein, C., S. Goldstone, and L. Lutzenhiser. 2000. 'A theory-based approach to market transformation.' *Energy Policy* 28:137–44.

Borenstein, S. 2002. 'The trouble with electricity markets: Understanding California's restructuring disaster.' *Journal of Economic Perspectives* 16:191–211.

California Energy Commission (CEC). 2003a. *Revised energy conservation impact assessment*. Sacramento, CA: California Energy Commission.

– 2003b. *California energy demand 2003-2013 forecast*. Staff draft report no. 100-03-02SD. Sacramento, CA: California Energy Commission.

– 2008. PLACE^{3}S (Planning for Community, Energy, Environmental, and Economic Sustainability). http://www.energy.ca.gov/places (accessed 9 September 2008).

California Institute for Energy and Environment (CIEE). 2007. Advisory Summit: Behavior, Energy, and Climate Change Project. Sacramento, CA. 31 May.

– 2008. Edward Vine, program manager, personal communication.

Cramer, J., N. Miller, P. Craig, B. Hackett, T. Dietz, M. Levine, and D. Kowalczyk. 1985. 'Social and engineering determinants and their equity implications in residential electricity use.' *Energy* 10:1283–91.

Energy Information Administration (EIA). 2003. 'Status of electric industry restructuring activity as of February 2003.' http://www.eia.doe.gov/cneaf/electricity/chg_str/restructure.pdf.

Geller, H., and S. Nadel. 1994. 'Market transformation strategies to promote end-use efficiency.' *Annual Review of Energy and the Environment* 19:301–46.

Goldman, C., J. Eto, and G. Barbose. 2002. *California customer load reductions during the electricity crisis: Did they help to keep the lights on?* LBNL 49733. Berkeley, CA: Lawrence Berkeley National Laboratory.

Harris, J., R. Diamond, M. Iyer, C. Payne, and C. Blumstein. 2006. 'Don't supersize me! Toward a policy of consumption-based energy efficiency.' In *Proceedings, American Council for an Energy Efficient Economy*, 7.100–7.114. Washington, DC: ACEEE Press.

Hirsh, R. 1999. 'PURPA: The spur to competition and utility restructuring.' *Electricity Journal* 12:60–72.

Hughes, T.P. 1989. 'The evolution of large technological systems.' In *The social construction of technological systems*, ed. W. Bijker, T.P. Hughes, and T. Pinch, 51–82. Cambridge, MA: MIT Press.

Intergovernmental Panel on Climate Change (IPCC). 2008. *Climate change 2007 – Impacts, adaptation, and vulnerability: Working Group II contribution to the fourth assessment report of the IPCC*. Cambridge: Cambridge University Press.

Kandel, A., M. Sheridan, and P. McAuliffe. 2008. 'A comparison of per capita electricity consumption in the United States and California.' In *Proceedings, American Council for an Energy Efficient Economy*, 8.123–8.134. Washington, DC: ACEEE Press.

Keirstead, J. 2006. 'Evaluating the applicability of integrated domestic energy consumption frameworks in the UK.' *Energy Policy* 34: 3065–77.

Kriström, B. 2006. 'Residential energy demand: A survey.' Presented at OECD workshop on sustainable consumption, Paris, 15-16 June.

Kunkle, R., L. Lutzenhiser, S. Sawyer, and S. Bender. 2004. 'New imagery and directions for residential sector energy policies.' In *Proceedings, American Council for an Energy Efficient Economy*, 7:171–82. Washington, DC: ACEEE Press.

Lovins, A. 1989. 'The negawatt revolution: Solving the CO2 problem.' Keynote address at the Green Energy Conference organized by the Canadian Coalition for Nuclear Responsibility, Montreal. http://www.ccnr.org/amory.html (accessed 9 September 2008).

Lutzenhiser, L. 1993. 'Social and behavioral aspects of energy use.' *Annual Review of Energy and the Environment* 18:247–89.

– 2002. 'Greening the economy from the bottom-up? Lessons in consumption from the energy case.' In *Readings in Economic Sociology*, ed. Nicole W. Biggart, 345–56. Oxford: Blackwell.

Lutzenhiser, L., and S. Bender. 2008. 'The "average American" unmasked: Social structure and differences in household energy use and carbon emissions.' In *Proceedings, American Council for an Energy Efficient Economy*, 7.191–7.204. Washington, DC: ACEEE Press.

Lutzenhiser, L., and B. Hackett. 1993. 'Social stratification and environmental degradation: Understanding household CO2 production.' *Social Problems* 40:50–73.

Lutzenhiser, L., R. Kunkle, J. Woods, and S. Lutzenhiser. 2003. 'Conservation behavior by residential consumers during and after the 2000–2001 California energy crisis.' In *Public Interest Energy Strategies Report*, staff report no. 100-03-012D, 154–207. Sacramento, CA: California Energy Commission.

Meier A., A.H. Rosenfeld, and J. Wright. 1982. 'Supply curves of conserved energy for California's residential sector.' *Energy* 7(4): 347–58.

Metroquest. 2008. 'Envision your future.' Envision Sustainability Tools. Vancouver, BC. http://www.metroquest.com (accessed 15 October 2009).

National Science Foundation (NSF). 2007. 'Dynamics of coupled natural and human systems (CNH): Program solicitation.' http://www.nsf.gov/pubs/2007/nsf07598/nsf07598.pdf (accessed 9 September 2008).

Pacala, S., and R. Socolow. 2004. 'Stabilization wedges: Solving the climate problem for the next 50 years with current technologies.' *Science* 305:968–72.

Princen, T. 2006. Keynote speech at the ACEEE 2006 Summer Study on Energy Use in Buildings. (Author of The logic of sufficiency. 2005. Cambridge, MA: MIT Press.)

Rudin, Andrew. 2000. 'Why we should change our message from "Use energy efficiently' to "Use less energy."' In *Proceedings, American Council for an Energy Efficient Economy*, 8.329–8.340. Washington, DC: ACEEE Press.

Rufo, M., and A. North. 2007. *Assessment of long-term electric energy efficiency potential in California's residential sector.* Itron, Inc. Consultant Report CEC-500-2007-002. Sacramento, CA: California Energy Commission.

Shove, E. 2004. *Comfort, cleanliness and convenience: The social organization of normality.* Oxford and New York: Berg Publishers.

Shove, E., L. Lutzenhiser, S. Guy, B. Hackett, and H. Wilhite. 1998. 'Energy and social systems.' In *Human choice and climate change*, ed. Steve Rayner and Elizabeth Malon, 201–34. Columbus, OH: Battelle Press.

Stern, P.C. 1986. 'Blind spots in policy analysis: What economics doesn't say about energy use.' *Journal of Policy Analysis and Management* 5:200–27.

Wilhite, H., and L. Lutzenhiser. 1999. 'Social loading and sustainable consumption.' *Advances in Consumer Research* 26:281–7.

Wilhite, H., E. Shove, L. Lutzenhiser, and W.Kempton. 2001. 'The legacy of twenty years of energy demand management: We know more about individual behavior but next to nothing about demand.' In *Society, behaviour and climate change mitigation*, ed. Ebarhard Jochem, Jayant Sathaye, and Daniel Bouille, 109–26. Dordrecht, Netherlands: Kluwer Academic Publishers.

Wilson, C., and H. Dowlatabadi. 2007. 'Models of decision making and residential energy use.' *Annual Review of Environment and Resources* 32:169–203.

13 A Review of the Rebound Effect in Energy Efficiency Programs

STEVE SORRELL

To achieve reductions in carbon emissions, most governments are seeking ways to improve energy efficiency throughout the economy. It is generally assumed that such improvements will reduce overall energy consumption, at least compared to a scenario in which such improvements are not made. However, a range of mechanisms, commonly grouped under the heading of *rebound effects*, may reduce the size of the energy savings achieved. Indeed, there is some evidence to suggest that the introduction of certain types of energy efficient technology in the past has contributed to an overall *increase* in energy demand – an outcome that has been termed *backfire*. This applies in particular to pervasive new technologies, such as steam engines in the nineteenth century, that significantly raise overall economic productivity as well as improve energy efficiency (Alcott 2005).

Rebound effects could have far-reaching implications for energy and climate policy. While cost-effective improvements in energy efficiency should improve welfare and benefit the economy, they could in some cases provide an ineffective or even a counterproductive means of tackling climate change. However, it does not necessarily follow that *all* improvements in energy efficiency will increase overall energy consumption or, in particular, that the improvements induced by policy measures will do so.

An earlier version of this chapter was published in *The International Handbook of the Economics of Energy* (Hunt and Evans 2008) and based upon a comprehensive review of the evidence for rebound effects, as conducted by the UK Energy Research Centre (Sorrell 2007).

The nature, operation, and importance of rebound effects are the focus of a long-running debate within energy economics. On the micro level, the question is whether improvements in the technical efficiency of energy use can be expected to reduce energy consumption by the amount predicted by simple engineering calculations. For example, will a 20% improvement in the fuel efficiency of passenger cars lead to a corresponding 20% reduction in motor-fuel consumption for personal automotive travel? Simple economic theory suggests that it will not. Since energy efficiency improvements reduce the marginal cost of energy services such as travel, the consumption of those services may be expected to increase. For example, since the cost per kilometre of driving is cheaper, consumers may choose to drive farther and/or more often. This increased consumption of energy services may be expected to offset some of the predicted reduction in energy consumption.

This so-called *direct rebound effect* was first brought to the attention of energy economists by Daniel Khazzoom (1980) and has since been the focus of much research (Greening, Greene, and Difiglio 2000). However, even if there is no direct rebound effect for a particular energy service (for example, even if consumers choose not to drive any farther in their fuel-efficient car), there are a number of other reasons that the economy-wide reduction in energy consumption may be less than simple calculations suggest. For example, the money saved on motor-fuel consumption may be spent on other goods and services that also require energy for their provision. These so-called *indirect rebound effects* can take a number of forms that are briefly outlined in box 13.1. Both direct and indirect rebound effects apply equally to energy efficiency improvements by consumers, such as the purchase of a more fuel-efficient car (figure 13.1), and energy efficiency improvements by producers, such as the adoption of energy efficient process technology (figure 13.2).

As shown in box 13.2, the overall or economy-wide rebound effect from an energy efficiency improvement represents the sum of these direct and indirect effects. Rebound effects are normally expressed as a percentage of the *expected* energy savings from an energy efficiency improvement. Hence, a rebound effect of 100% means that the expected energy savings are entirely offset, leading to zero net savings.[1] Backfire means that rebound effects exceed 100%.

Rebound effects need to be defined in relation to a particular time frame (for example, short-, medium-, or long-term) and a system boundary for the relevant energy consumption (for example, house-

Box 13.1. Indirect Rebound Effects

- The equipment used to improve energy efficiency (e.g., thermal insulation) will itself require energy to manufacture and install, and this 'embodied' energy consumption will offset some of the energy savings achieved.
- Consumers may use the cost savings from energy efficiency improvements to purchase other goods and services that themselves require energy to provide. For example, the cost savings from a more energy efficient central heating system may be put towards an overseas holiday.
- Producers may use the cost savings from energy efficiency improvements to increase output, thereby increasing consumption of capital, labour, and materials inputs that themselves require energy to provide. If the energy efficiency improvements are sector wide, they may lead to lower product prices, increased consumption of the relevant products, and further increases in energy consumption.
- Cost-effective energy efficiency improvements will increase the overall productivity of the economy, thereby encouraging economic growth. The increased consumption of goods and services may in turn drive up energy consumption.
- Large-scale reductions in energy demand may translate into lower energy prices that will encourage energy consumption to increase. The reduction in energy prices will also increase real income, thereby encouraging investment and generating an extra stimulus to aggregate output and energy use.
- Both the energy efficiency improvements and the associated reductions in energy prices will reduce the cost of energy intensive goods and services to a greater extent than that of non-energy intensive goods and services, thereby encouraging consumer demand to shift towards the former.

hold, firm, sector, national economy). The economy-wide effect is normally defined in relation to a national economy, but there may also be effects in other countries through changes in trade patterns and international energy prices. Rebound effects may also be expected to increase in importance over time as markets, technology, and behaviour adjust.

Figure 13.1. Illustration of Rebound Effects for Consumers

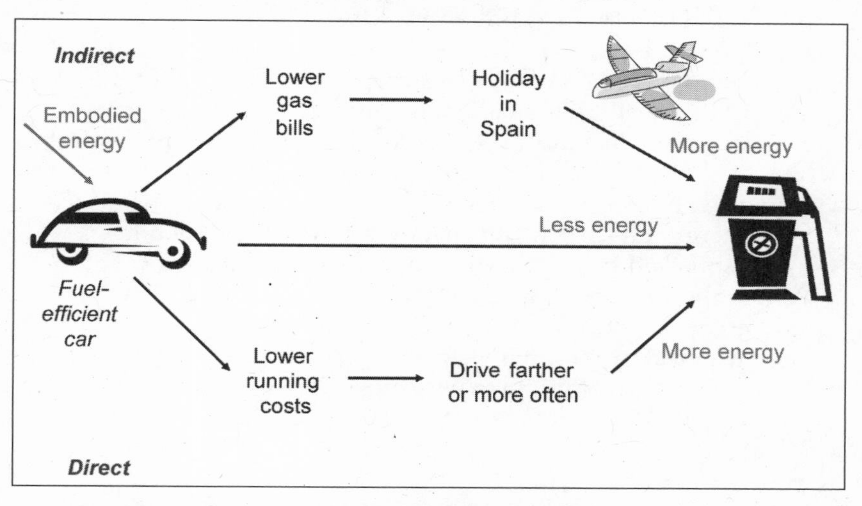

Figure 13.2. Illustration of Rebound Effects for Producers

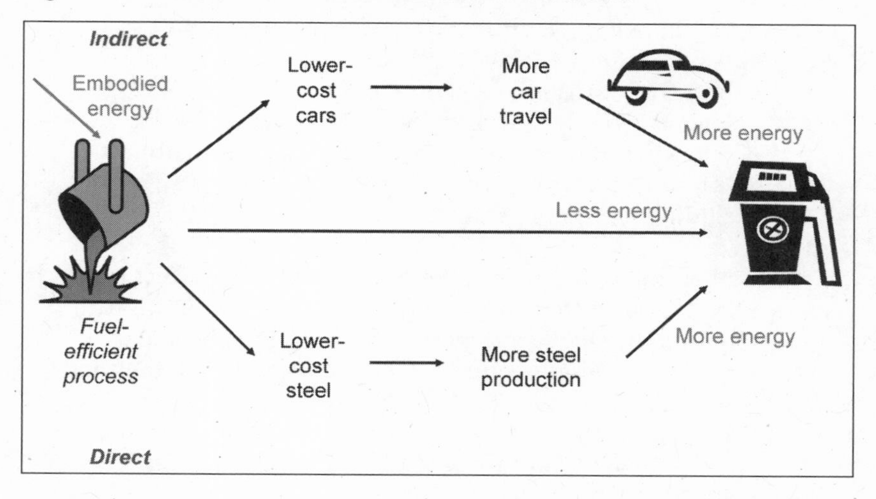

Rebound effects tend to be almost universally ignored in official analyses of the potential energy savings from energy efficiency improvements. A rare exception is the United Kingdom policy to improve the thermal insulation of housing, where it is expected that some of the benefits will be taken as higher internal temperatures rather than

Box 13.2. Classifying Rebound Effects

The economy-wide rebound effect represents the sum of the direct and indirect effects. For energy efficiency improvements by consumers, it is helpful to decompose the **direct rebound effect** into

a) a *substitution effect*, whereby consumption of the (cheaper) energy service substitutes for the consumption of other goods and services while maintaining a constant level of 'utility,' or consumer satisfaction; and
b) an *income effect*, whereby the increase in real income achieved by the energy efficiency improvement allows a higher level of utility to be achieved by increasing consumption of all goods and services, including the energy service.

Similarly, the direct rebound effect for producers may be decomposed into

a) a *substitution effect*, whereby the cheaper energy service substitutes for the use of capital, labour, and materials in producing a constant level of output; and
b) an *output effect*, whereby the cost savings from the energy efficiency improvement allow a higher level of output to be produced, thereby increasing consumption of all inputs, including the energy service.

It is also helpful to decompose the **indirect rebound effect** into

a) the *embodied energy*, or the indirect energy consumption required to achieve the energy efficiency improvement, such as the energy required to produce and install thermal insulation; and
b) the *secondary effects* that result as a consequence of the energy efficiency improvement, which include the mechanisms listed in box 13.1.

A diagrammatic representation of this classification scheme is provided below. The relative size of each effect may vary widely from one circumstance to another, and in some cases individual-components of the rebound effect may be negative. For example, if an energy service is an 'inferior good,' the income effect for consumers may lead to reduced consumption of that service, rather than increased consumption.

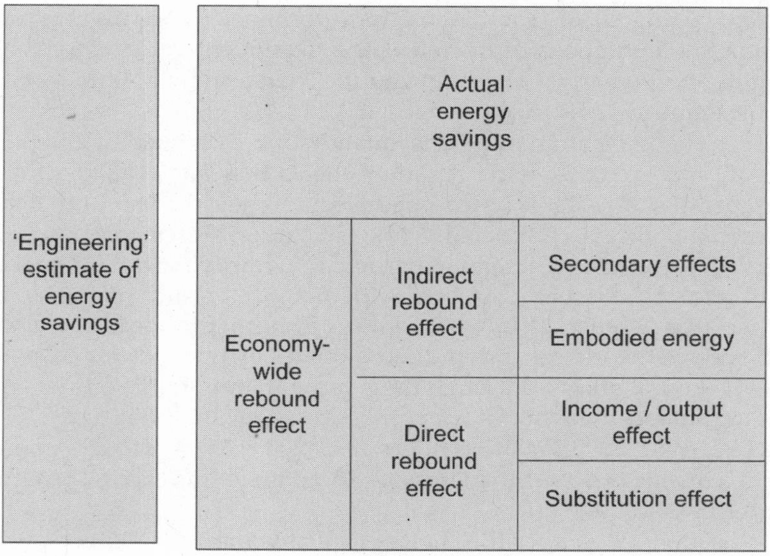

'Engineering' estimate of energy savings

Actual energy savings

Economy-wide rebound effect

Indirect rebound effect

Direct rebound effect

Secondary effects

Embodied energy

Income / output effect

Substitution effect

reduced energy consumption (UK Department of Environment, Food, and Rural Affairs 2007). However, the direct rebound effects for other energy efficiency measures are generally ignored, as are the potential indirect effects for all measures. Much the same applies to energy modelling studies and to independent estimates of energy efficiency potentials. For example, the Stern review of the economics of climate change overlooks rebound effects altogether (Stern 2007), while the fourth assessment report from the Intergovernmental Panel on Climate Change simply notes that the literature is divided on the magnitude of the effect (IPCC 2007).

While energy economists recognize that rebound effects may reduce the energy savings from energy efficiency improvements, there is dispute over how important these effects are. Some argue that rebound effects are of minor importance for most energy services, largely because the demand for those services appears to be inelastic in most cases and because energy typically forms a small share of the total costs of those services (Lovins 1998; Lovins et al. 1988; Schipper and Grubb 2000). Others argue that they are sufficiently important to completely offset the energy savings from improved energy efficiency (L.G. Brookes 2000; Herring 2006). The dispute has a number of origins, including competing definitions of the relevant independent variable for the rebound effect (energy efficiency) and the sparse and ambiguous nature of the empirical evidence (Sorrell 2007).

This chapter summarizes the quantitative estimates of direct, indirect, and economy-wide rebound effects that are available from a number of sources. It is based upon a comprehensive literature review of rebound effects, conducted by the UK Energy Research Centre (Sorrell 2007). The review sought in particular to clarify the definitional and methodological issues associated with quantifying such effects, an emphasis that is reflected in what follows. This chapter does not discuss the various theoretical and 'indirect' sources of evidence for economy-wide rebound effects, although these form an important component of the arguments in favour of backfire (Alcott 2005; L. Brookes 2004; L.G. Brookes 1984, 1990, 2000; Saunders 1992, 2000, 2007). Readers interested in a review of these broader issues should refer instead to Sorrell and Dimitropoulos (2007d).

Section 2 describes the choices available for the independent and dependent variables for the rebound effect and the possible implications of these choices. Section 3 describes the nature and operation of direct rebound effects, summarizes the quantitative estimates that are available from a number of studies, and highlights some potential sources of bias. Section 4 describes the nature and operation of indirect and economy-wide rebound effects, summarizes the quantitative estimates available from energy modelling studies, and highlights some potential methodological weaknesses. Section 5 highlights some policy implications of rebound effects, whilst Section 6 concludes.

Definitional Issues

Energy efficiency improvements are generally assumed to reduce energy consumption below where it would have been without those improvements. Rebound effects reduce the size of these energy savings. However, estimating the size of any energy savings is far from straightforward because (1) real-world economies do not permit controlled experiments, so the relationship between a change in energy efficiency and a subsequent change in energy consumption is likely to be mediated by a host of confounding variables; (2) we cannot observe what energy consumption would have been without the energy efficiency improvement, so the estimated savings will always be uncertain; and (3) energy efficiency is not controlled externally by an experimenter and may be influenced by a variety of technical, economic, and policy variables. In particular, the direction of causality may run in reverse, with changes in energy consumption (whatever their cause) leading to changes in different measures of energy efficiency.

Energy efficiency may be defined as the ratio of useful outputs to energy inputs for a system. The system in question may be an individual energy conversion device (for example, a boiler), a building, an industrial process, a firm, a sector, or an entire economy. In all cases, the measure of energy efficiency will depend upon how *useful* is defined and how inputs and outputs are measured. The options include *thermodynamic measures,* where the outputs are defined in terms of heat content or the capacity to perform useful work; *physical measures,* where the outputs are defined in physical terms, such as vehicle kilometres or tonnes of steel; or *economic measures,* where the outputs are defined in economic terms, such as value-added or gross domestic product (GDP). Economic measures of energy efficiency are frequently referred to as *energy productivity*.

The choice of measures for inputs and outputs, the appropriate system boundaries, and the time frame under consideration can vary widely from one study to another. The conclusions drawn regarding the magnitude and importance of rebound effects will depend upon the particular choices that are made.

Economists are primarily interested in energy efficiency improvements that are consistent with the best use of all economic resources. These are conventionally divided into two categories: those that are associated with improvements in overall, or 'total factor,' productivity (*technical change*); and those that are not (*substitution*). The former are usually assumed to occur independently of changes in relative prices, while the latter are assumed to occur in response to such changes. The consequences of technical change are of particular interest as this contributes to the growth in economic output. However, distinguishing empirically between the two can be challenging.

Many commentators assume that the relevant independent variable for the rebound effect is improvements in the thermodynamic efficiency of individual conversion devices or industrial processes. However, such improvements will only translate into comparable improvements in different measures of energy efficiency, or measures of energy efficiency applicable to wider system boundaries, if several of the mechanisms responsible for the rebound effect fail to come into play. For example, improvements in the number of litres used per vehicle-kilometre will only translate into improvements in the number of litres used per passenger-kilometre if there are no associated changes in average vehicle load factors.

Rebound effects may be expected to increase over time and with the widening of the system boundary for the dependent variable (energy consumption). Hence, to capture the full range of rebound effects, the system boundary for the independent variable (energy efficiency) should be relatively narrow, while the system boundary for the dependent variable should be as wide as possible. However, estimating the economy-wide effects of micro-level improvements in the thermodynamic efficiency is, at best, challenging. Partly for this reason, the independent variable for many theoretical and empirical studies of rebound effects is a physical or economic measure of energy efficiency that is applicable to relatively wide system boundaries – such as the energy efficiency of an industrial sector. But such studies may overlook the 'lower-level' rebound effects resulting from improvements in physical or thermodynamic measures of energy efficiency appropriate to narrower system boundaries (for example, the diffusion of energy efficient motors within the sector).[2] Also, improvements in aggregate measures of energy efficiency are unlikely to be caused solely (or even mainly) by the diffusion of more thermodynamically efficient conversion devices.

Aggregate measures of energy efficiency will also depend upon how different types of energy input are combined. While it is common practice to aggregate different energy types on the basis of their heat content, this neglects the thermodynamic quality of each energy type, or its ability to perform useful work.[3] That ability, in turn, is only one of several factors that determine the economic productivity of different energy types, with others including cleanliness, amenability to storage, safety, flexibility, and the use to which the energy is put (Cleveland, Kaufmann, and Stern 2000).[4] In general, when the quality of energy inputs are accounted for, aggregate measures of energy efficiency are found to be improving more slowly than is commonly supposed (Cleveland, Kaufmann, and Stern 2000; Hong 1983; Zarnikau 1999).[5]

Improvements in any measure of energy efficiency rarely occur in isolation but are typically associated with broader improvements in the productivity of other inputs. As illustrated in box 13.3, this may be the case even when the primary objective of the relevant investment is to reduce energy costs. Importantly, if the total cost savings exceed the saving in energy costs, then any rebound effects may be amplified.

Evidence for Direct Rebound Effects

This section summarizes the empirical evidence for direct rebound effects, focusing in particular on energy services in the household sector because this is where the bulk of evidence lies.

Direct rebound effects relate to individual energy services, such as heating, lighting, and refrigeration, and are confined to the energy required to provide that service. Such services are provided through a combination of capital equipment, labour, materials, and energy. An essential feature of an energy service is the *useful work* obtained – with the term being used here to refer to both thermodynamic and physical measures of useful outputs. Energy services may also have broader attributes that may be combined with useful work in a variety of ways. For example, all cars deliver passenger-kilometres, but they may vary widely in terms of speed, comfort, acceleration, and prestige. Consumers and producers may therefore make trade-offs between useful work and other attributes of an energy service; between energy, capital, and other market goods in the production of an energy service; and between different types of energy service.

By reducing the marginal cost of useful work, energy efficiency improvements may, over time, lead to an increase in the number of energy conversion devices, their average size, their average utilization, and/or their average load factor. For example, people may buy more cars, buy larger cars, drive them farther, and/or share them less. The relative importance of these variables may be expected to vary widely between different energy services and over time. Over the very long term, the lower cost of energy services may contribute to fundamental changes in technologies, infrastructures, and lifestyles – such as a shift towards car-based commuting and increasing distances between residential, workplace, and retail locations. But as the time horizon extends, the effect of such changes on the demand for the energy service becomes increasingly difficult to separate from the effect of income growth and other factors.

Box 13.3. Examples of the Link Between Improved Energy Efficiency and Improved Total Factor Productivity

- Lovins and Lovins (1997) used case studies to argue that better visual, acoustic, and thermal comfort in well-designed, energy efficient buildings can improve labour productivity by as much as 16%. Since labour costs in commercial buildings are typically twenty-five times greater than energy costs, the resulting cost savings can potentially dwarf those from reduced energy consumption.
- Pye and McKane (1998) showed how the installation of energy efficient motors reduced wear and tear, extended the lifetime of system components, and achieved savings in capital and labour costs that exceeded the reduction in energy costs.
- Worrell et al. (2003) analyzed the cost savings from fifty-two energy efficiency projects, including motor replacements, fans/duct/pipe insulation, improved controls, and heat recovery in a range of industrial sectors. The average payback period from energy savings alone was 4.2 years, but this fell to 1.9 years when the non-energy benefits were taken into account.
- Using plant-level data, Boyd and Pang (2000) estimated fuel and electricity intensity in the glass industry as a function of energy prices, cumulative output, a time trend, capacity utilization, and overall productivity. Their results show that the most productive plants are also most energy efficient and that a 1% improvement in overall productivity results in a more than 1% improvement in energy efficiency.

The estimated size of the direct rebound effect will depend upon how useful work, and hence energy efficiency, is defined. For example, the majority of estimates of the direct rebound effect for personal automotive transport measure useful work in terms of vehicle-kilometres, which is sometimes decomposed into the product of the number of vehicles and the mean distance travelled per vehicle per year. Energy efficiency is then defined as vehicle-kilometres per litre of fuel, and direct rebound effects are measured as increases in distance driven. However, this overlooks any changes in mean vehicle size and weight as a result of energy efficiency improvements (for example, more SUVs), as well as any decrease in average vehicle load factor (for example, less car sharing).[6]

The magnitude of direct rebound effects may be expected to be proportional to the share of energy in the total cost of energy services,[7] as well as the extent to which those costs are visible. However, as the consumption of a particular energy service increases, saturation effects (technically, declining marginal utility) should reduce the size of any direct rebound effect. This suggests that direct rebound effects will be higher among low income groups because these are further from satiation in their consumption of many energy services (Milne and Boardman 2000).

Increases in demand for an energy service may derive from existing consumers of the service or from consumers who were previously unable or unwilling to purchase that service. For example, improvements in the energy efficiency of home air conditioners may encourage consumers to purchase portable air conditioners for the first time. The abundance of such 'marginal consumers' (Wirl 1997) in developing countries points to the possibility of large direct rebound effects in these contexts, offset to only a limited extent by saturation effects among existing consumers (Roy 2000).

While energy efficiency improvements reduce the energy cost of useful work, the size of the direct rebound effect will depend upon how other costs are affected. For example, direct rebound effects may be smaller if energy efficient equipment is more expensive than less efficient alternatives, because the availability of such equipment should not encourage an increase in the number and capacity of the relevant conversion devices. However, once purchased, such equipment may be expected to have a higher utilization. In practice, many types of equipment appear to have both improved in energy efficiency over time and fallen in total cost relative to income.

Even if energy efficiency improvements are not associated with changes in capital or other costs, certain types of direct rebound effect may be constrained by the real or opportunity costs associated with increasing demand. Two examples are the opportunity cost of space (for example, increasing refrigerator size may not be the best use of available space) and the opportunity cost of time (for example, driving longer distances may not be the best use of available time). However, space constraints may become less important if technological improvements reduce the average size of conversion devices or if rising incomes lead to an increase in average living space (Wilson and Boehland 2005). In contrast, the opportunity cost of time should increase with income.

Approaches to Estimating Direct Rebound Effects

There are two broad approaches to estimating direct rebound effects, namely quasi-experimental and econometric.

The Quasi-Experimental Approach. One approach relies on measuring the demand for useful work before and after an energy efficiency improvement; for example, measuring the change in heat output following the installation of a fuel-efficient boiler. The demand for useful work before the energy efficiency improvement could be taken as an estimate for what demand would have been in the absence of the improvement. However, various other factors may also have changed the demand for useful work, which need to be controlled (Frondel and Schmidt 2005; Meyer 1995).

Since it can be very difficult to measure useful work for many energy services, an alternative approach is to measure the change in energy consumption for that service. In order to estimate direct rebound effects, this needs to be compared with a counterfactual estimate of energy consumption that has at least two sources of error: (1) the energy consumption that would have occurred without the energy efficiency improvement; and (2) the energy consumption that would have occurred following the energy efficiency improvement, had there been no behavioural change. The first of these gives an estimate of the energy savings from the energy efficiency improvement, while the second isolates the direct rebound effect. Estimates for the latter can be derived from engineering models, but these frequently require data on the circumstances of individual installations and are prone to error.

Both of these approaches are rare, owing in part to measurement difficulties. There are relatively few published studies, and nearly all of these focus on household heating (Sommerville and Sorrell 2007). The methodological quality of most of these studies is relatively poor, with the majority using simple before and after comparisons, without the use of a control group or explicitly controlling for confounding variables. This is the weakest methodological strategy and prone to bias (Frondel and Schmidt 2005; Meyer 1995). Also, several studies are vulnerable to selection bias because households choose to participate rather being randomly assigned (Hartman 1988). Other weaknesses include small sample sizes, high variance in results, and monitoring periods that are too short to capture long-term changes. There is also confusion between (1) *shortfall*, the difference between actual savings in energy consumption and those expected on the basis of engineering estimates; (2) *temperature take-back*, the change in mean internal temperatures following the energy efficiency improvement and/or the associated reduction in energy savings; and (3) *behavioural change*, the proportion of the change in internal temperature that derives from adjustments of heating controls and other variables by the user (for example, opening windows) and/or the associated reduction in energy savings.

Typically, only a portion of temperature take-back is due to behavioural change, with the remainder being due to physical and other factors (Sanders and Phillipson 2006).[8] Similarly, only a portion of shortfall is due to temperature take-back, with the remainder being due to poor engineering estimates of potential savings, inadequate performance of equipment, deficiencies in installation, and so on. Hence, behavioural change is one, but not the only (or necessarily the most important), explanation of temperature take-back; and the latter is one, but not the only explanation, of shortfall. Direct rebound effects are normally interpreted as behavioural change, but it may be misleading to interpret this solely as a rational response to lower heating costs, partly because energy efficiency improvements may change other variables (for example, airflow) that also encourage behavioural responses. Also, measures of temperature take-back may be difficult to translate into estimates of shortfall because of a non-linear and household-specific relationship between energy consumption and internal temperature. Isolating rebound effects from such studies can therefore be challenging.

The Econometric Approach. A more common approach to estimating direct rebound effects is through the econometric analysis of secondary data sources that include information on the demand for energy, useful work, and/or energy efficiency. This data can take a number of forms (for example, cross-sectional, time-series) and apply to the household, regional, or country level. Such studies typically estimate *elasticities*, meaning the percentage change in one variable following a percentage change in another, holding other variables constant. If time-series data is available, an estimate can be made of short-run elasticities, where the stock of conversion devices is assumed to be fixed, as well as long-run elasticities, where it is variable. Cross-sectional data is usually assumed to provide estimates of long-run elasticities.

Depending upon data availability, the direct rebound effect may be estimated from one of two *energy efficiency* elasticities:[9]

- $\eta_\varepsilon(E)$: the elasticity of demand for energy (E) with respect to energy efficiency (ε)
- $\eta_\varepsilon(S)$: the elasticity of demand for useful work (S) with respect to energy efficiency (where $S = \varepsilon E$)

$\eta_\varepsilon(S)$ is generally taken as a direct measure of the rebound effect. Under certain assumptions, it can be shown that: $\eta_\varepsilon(E) = \eta_\varepsilon(S) - 1$ (Sorrell and Dimitropoulos 2007a). Hence, the actual saving in energy consumption will only equal the predicted saving from engineering calculations when the demand for useful work remains unchanged following an energy efficiency improvement (that is, when $\eta_\varepsilon(S) = 0$).[10]

Instead of using $\eta_\varepsilon(E)$ or $\eta_\varepsilon(S)$, most studies estimate the rebound effect from one of three *price* elasticities:

- $\eta_{P_S}(S)$: the elasticity of demand for useful work with respect to the energy cost of useful work (P_S);
- $\eta_{P_E}(S)$: the elasticity of demand for useful work with respect to the price of energy;
- $\eta_{P_E}(E)$: the elasticity of demand for energy with respect to the price of energy;

where $P_S = P_E / \varepsilon$. Under certain assumptions, the negative of either $\eta_{P_S}(S)$, $\eta_{P_E}(S)$ or $\eta_{P_E}(E)$ can be taken as an approximation to $\eta_\varepsilon(S)$ and hence may be used as a measure of the direct rebound effect (Sorrell and Dimitropoulos 2007a). The use of price elasticities in this way implicitly equates the direct rebound effect to a behavioural response to the lower cost of energy services. It therefore ignores any other reasons why the demand for useful work may change following an improvement in energy efficiency.

The choice of the appropriate elasticity measure will depend in part upon data availability. Generally, data on energy consumption (E) and energy prices (P_E) are both more available and more accurate than data on useful work (S) and energy efficiency (ε). Also, even if data on energy efficiency are available, the amount of variation is typically limited, with the result that estimates of either $\eta_\varepsilon(E)$ or $\eta_\varepsilon(S)$ can have a large variance. In contrast, estimates of $\eta_{P_S}(S)$ may have less variance owing to significantly greater variation in the independent variable. This is because the energy cost of useful work depends upon the ratio of energy prices to energy efficiency ($P_S = P_E / \varepsilon$), and most data sets include considerable cross-sectional or longitudinal variation in energy prices. In principle, rational consumers should respond in the same way to a decrease in energy prices as they do to an improvement in energy efficiency (and vice versa), as these should have an identical effect on the energy cost of useful work (P_S). However, there may be a number of reasons this 'symmetry' assumption does not hold. If so, estimates of the direct rebound effect that are based upon $\eta_{P_S}(S)$ could be biased.

Estimates of $\eta_{P_S}(S)$ are largely confined to personal automotive transportation, household heating, and space cooling where proxy measures of useful work are most readily available. These energy services form a significant component of household energy demand in OECD countries and may be expected to be relatively price elastic. There are very few estimates of $\eta_{P_S}(S)$ for other consumer energy services and practically none for producers. Furthermore, the great majority of studies refer to the United States.

In many cases, data on energy efficiency are either unavailable or inaccurate. In these circumstances, the direct rebound effect may be estimated from $\eta_{P_E}(S)$ and $\eta_{P_E}(E)$, but this is only valid if, first, consumers respond in the same way to a decrease in energy prices as they do to an improvement in energy efficiency (and vice versa) and, second, energy efficiency is unaffected by changes in energy prices. Both these assumptions are likely to be flawed, but the extent to which this leads to biased estimates of the direct rebound effect may vary widely from one energy service to another and between the short and long term.

Under certain assumptions, the own-price elasticity of energy demand ($\eta_{P_E}(E)$) for a particular energy service can be shown to provide an *upper bound* for the direct rebound effect (Barker, Ekins, and Johnstone 1995; Dahl 1993, 1994; Dahl and Sterner 1991; Espey and Espey 2004; Espey 1998; Graham and Glaister 2002; Hanley, Dargay, and Goodwin 2002). Reviews of the extensive literature on this topic generally suggest that energy demand is inelastic in the majority of sectors in OECD countries (that is, $|\eta_{P_E}(E)| < 1$) (Barker, Ekins, and Johnstone 1995). The implication is that the direct rebound effect alone is unlikely to lead to backfire in these circumstances – although there are undoubtedly exceptions.

For the purpose of estimating rebound effects, estimates of $\eta_{P_E}(E)$ are most useful when the energy demand in question relates to a single energy service, such as refrigeration. They are less useful when (as is more common) the measured demand derives from a collection of energy services, such as household fuel or electricity consumption. In this case, a large value for $\eta_{P_E}(E)$ may suggest that improvements in the overall efficiency of fuel or electricity use will lead to large direct rebound effects (and vice versa) or that the direct rebound effect for the energy services that dominate fuel or electricity consumption may be large. However, a small value for $\eta_{P_E}(E)$ would not rule out the possibility of large direct rebound effects for individual energy services.

Whatever their scope and origin, estimates of price elasticities should be treated with caution. Aside from the difficulties of estimation, behavioural responses are contingent upon technical, institutional, policy, and demographic factors that vary widely between different groups and over time. Demand responses are known to vary with the level of prices, the origin of price changes (for example, exogenous versus policy-induced), expectations of future prices, government fiscal policy (for example, recycling of carbon tax revenues), saturation effects, and other factors (Sorrell and Dimitropoulos, 2007c). The past is not necessarily a good guide to the future in this area, and it is possible that the very long-run response to price changes may exceed those found in empirical studies that rely upon data from relatively short time periods.

Estimates of Direct Rebound Effects

By far the best studied area for the direct rebound effect is personal automotive transport. Most studies refer to the United States, which is important because fuel prices, fuel efficiencies, and residential densities are lower than in Europe, car ownership levels are higher, and there is less scope for switching to alternative transport modes.

Studies estimating $\eta_\varepsilon(E)$, $\eta_\varepsilon(S)$, or $\eta_{P_S}(S)$ for personal transport vary considerably in terms of the data used and the specifications employed. Most studies use aggregate data that can capture long-term effects on demand such as fuel efficiency standards, while household survey data can better describe individual behaviour at the micro level. Aggregate studies face numerous measurement difficulties, however (Schipper, Josefina, Figueroa, and Espey 1993), while disaggregate studies produce results that are more difficult to generalize. While all of these studies use distance travelled as a measure of useful work (S), this may either be measured in absolute terms or normalized to the number of adults, licensed drivers, households, or vehicles (Sorrell and Dimitropoulos 2007b). The relevant estimates may be expected to differ as a result.

Following a review of seventeen econometric studies of personal automotive transport, Sorrell and Dimitropoulos (2007c) concludes that the long-run direct rebound effect lies somewhere between 10% and 30%. The most reliable estimates come from studies using aggregate panel data, owing to the greater number of observations. For example, Johansson and Schipper's (1997) cross-country study gives a best guess for the long-run direct rebound effect of 30%, while both Haughton and Sarkar (1996) and Small and Van Dender (2005) converge on a long-run value of 22% for the United States (see box 13.4). Most studies assume that the response to a change in fuel prices is equal in size to the response to a change in fuel efficiency but opposite in sign. However, few studies test this assumption explicitly, and those that do are either unable to reject the hypothesis that the two elasticities are equal or find that the fuel efficiency elasticity is *less* than the fuel cost per kilometre elasticity. The implication is that the direct rebound effect may lie towards the lower end of the 10%–30% range.

A number of studies suggest that the direct rebound effect for personal automotive travel declines with income, as theory predicts (Greene, Kahn, and Gibson 1999; Small and Van Dender 2005). The evidence is insufficient to determine whether direct rebound effects are larger or smaller in Europe, but it is notable that the meta-analysis by Espey (1998) found no significant difference in long-run own-price elasticities of gasoline demand. Overall, Sorrell and Dimitropoulos (2007c) concludes that direct rebound effects in this sector have not obviated the benefits of technical improvements in vehicle fuel efficiency. Between 70% and 100% of the potential benefits of such improvements appear to have been realized in reduced consumption of motor fuels.

The next best studied area for direct rebound effects is household heating. Sommerville and Sorrell (2007) reviews fifteen quasi-experimental studies of this energy service and conclude that standard engineering models may overestimate energy savings by up to one-half – and potentially by more than this for low income households. However, overall percent shortfall is highly contingent on the accuracy of the engineering models, and attempts to calibrate these models to specific household conditions generally result in a lower shortfall.

Box 13.4. The Declining Direct Rebound Effect

Small and Van Dender (2005) provide one of the most methodologically rigorous estimates of the direct rebound effect for personal automotive transport. They estimate an econometric model, explaining the amount of travel by passenger cars as a function of the cost per mile and other variables. By employing simultaneous equations for vehicle numbers, average fuel efficiency, and vehicle miles travelled, they are able allow for the fact that fuel efficiency is endogenous; i.e., more fuel-efficient cars may encourage more driving, while the expectation of more driving may encourage the purchase of more fuel-efficient cars. Their results show that failing to allow for this can lead to overestimation of the direct rebound effect.

Small and Van Dender use aggregate data on vehicle numbers, fuel efficiency, gasoline consumption, vehicle miles travelled, and other variables for fifty U.S. states and the District of Columbia, covering the period of 1961 to 2001. This approach provides considerably more observations than do conventional aggregate time-series data, while at the same time providing more information on effects that are of interest to policymakers than do studies using household survey data. The effect of the Corporate Average Fuel Economy (CAFE) standards on vehicle fuel efficiency is estimated by incorporating a variable representing the gap between the fuel efficiency standard and an estimate of the efficiency that would have been chosen in the absence of the standards, given prevailing fuel prices.

Small and Van Dender estimate the short-run direct rebound effect for the United States as a whole to be 4.5% and the long-run effect to be 22%. The former is lower than most of the estimates in the literature, while the latter is close to the consensus. However, they estimate that a 10% increase in income reduces the short-run direct rebound effect by 0.58%. Using U.S. average values of income, urbanization and fuel prices over the period of 1997 to 2001, they find a direct rebound effect of only 2.2% in the short term and 10.7% in the long term – approximately half the values estimated from the full data set. If this result is robust, it has some important implications. However, two-fifths of the estimated reduction in the rebound effect derives from the assumption that the magnitude of this effect depends upon the absolute level of fuel costs per kilometre. However, since the relevant coefficient is not statistically significant, this claim is questionable.

Although methodologically sophisticated, the study is not without its problems. Despite covering fifty states over a period of thirty-six years, the data provide relatively little variation in vehicle fuel efficiency, making it difficult to determine its effect separately from the effect of fuel prices. Direct estimates of are small and statistically insignificant, which could be interpreted as implying that the direct rebound effect is approximately zero, but since this specification performs rather poorly overall, estimates based upon are preferred. Also, the model leads to the unlikely result that the direct rebound effect is negative in some states. This raises questions about the use of the model for projecting declining rebound effects in the future because increasing incomes could make the estimated direct rebound effect negative in many states (Harrison et al. 2005).

The studies provide mean estimates of temperature take-back in the range 0.14°C to 1.6°C, of which approximately half is estimated to be accounted for by the physical characteristics of the house and the remainder by behavioural change. Estimates of the energy savings lost through temperature take-back range from 0% to 100%, but with a mean around 20%. Temperature take-back appears to be higher for low income groups and for households with low internal temperatures prior to the efficiency measures. However, these two explanatory variables are likely to be correlated. As pre-intervention room temperatures approach 21°C, the magnitude of temperature take-back decreases owing to saturation effects.

Overall, while shortfall may often exceed 50% (especially for low income households), temperature take-back only accounts for a portion of this shortfall, and behavioural change only accounts for a portion of the temperature take-back. Temperature take-back would appear to reduce energy savings by around 20% on average, with the contribution from behavioural change being somewhat less. Which of these measures best corresponds to the direct rebound effect is a matter of debate.

Relatively few econometric studies estimate $\eta_{\varepsilon}(E)$, $\eta_{\varepsilon}(S)$, or $\eta_{P_S}(S)$ for household heating, and even fewer investigate rebound effects. Most studies rely upon detailed household survey data and exhibit considerable diversity in terms of the variables measured and the methodologies adopted. Sorrell and Dimitropoulos (2007c) reviews nine estimates and finds values in the range of 10% to 58% for the short run and 1.4% to 60% for the long run. As with the quasi-experimental studies, the definition of the direct rebound effect is not consistent between studies, and the behavioural response appears to vary widely between different households. Nevertheless, for the purpose of policy evaluation, a figure of 30% would appear to be a reasonable assumption.

Sorrell and Dimitropoulos (2007c) found only two studies of direct rebound effects for household cooling, and these provided estimates comparable to those for household heating (that is, 1% to 26%). However, these were relatively old studies, conducted during a period of rising energy prices and using small sample sizes. Their results may not be transferable to other geographical areas, owing to differences in house types and climatological conditions. Also, both studies focused solely upon changes in equipment utilization. To the extent that ownership of cooling technology is rapidly increasing in many countries, demand from 'marginal consumers' may be an important consideration, together with increases in system capacity among existing users.

Sorrell and Dimitropoulos (2007c) finds that the evidence for water heating is even more limited, although Guertin, Kumbhakar, and Duraiappah (2003) provides estimates in the range of 34% to 38%, which is significantly larger than the results from quasi-experimental studies reported by Nadel (1993). A methodologically rigorous study of direct rebound effects for clothes washing (box 13.5) suggests that direct rebound effects for 'minor' energy services should be relatively small (as theory suggests). However, this study confines attention to households that already have automatic washing machines, and therefore excludes rebound effects from marginal consumers.

Table 13.1 summarizes the results of Sorrell and Dimitropoulos's survey of econometric estimates of the direct rebound effect. Despite the methodological diversity, the results for individual energy services are broadly comparable, suggesting that the evidence is relatively robust to different data sets and methodologies. Also, consideration of the potential sources of bias (box 13.6) suggests that direct rebound effects are more likely to lie towards the lower end of the range indicated here. The results suggest that the mean long-run direct rebound effect for personal automotive transport, household heating, and household cooling in OECD countries is likely to be 30% or less and may

Box 13.5. Direct Rebound Effects for Clothes Washing

Davis (2007) provides a unique example of an estimate of direct rebound effects for household clothes washing, which together with clothes drying accounts for around one-tenth of U.S. household energy consumption. The estimate is based upon a government-sponsored field trial of high-efficiency washing machines involving ninety-eight participants. These machines use 48% less energy per wash than do standard machines, and 41% less water.

While participation in the trial was voluntary, both the utilization of existing machines and the associated consumption of energy and water were monitored for a period of two months prior to the installation of the new machine. This allowed household-specific variations in utilization patterns to be controlled for, and permitted unbiased estimates to be made of the price elasticity of machine utilization.

The monitoring allowed the marginal cost of clothes washing for each household to be estimated. This was then used as the primary independent variable in an equation for the demand for clean clothes in kg/day (useful work). Davis found that the demand for clean clothes increased by 5.6% after receiving the new washers, largely as a result of increases in the weight of clothes washed per cycle rather than in the number of cycles. While this could be used as an estimate of the direct rebound effect, it results in part from savings in water and detergent costs. If the estimate were based solely on the savings in energy costs, the estimated effect would be smaller. This suggests that only a small portion of the gains from energy efficient washing machines will be offset by increased utilization.

Davis estimates that time costs form 80%–90% of the total cost of washing clothes. The results therefore support the theoretical prediction that, for time-intensive activities, even relatively large changes in energy efficiency should have little impact on demand. Similar conclusions should therefore apply to other time-intensive energy services that are both produced and consumed by households, including those provided by dishwashers, vacuum cleaners, televisions, power tools, computers, and printers.

Box 13.6. Potential Sources of Bias in Estimates of the Direct Rebound Effect

Most estimates of the direct rebound effect assume that changes in energy prices have an opposite effect to changes in energy efficiency and that the latter are exogenous. In practice, both of these assumptions may be incorrect.

First, while changes in energy prices are generally not correlated with changes in other input costs, changes in energy efficiency may be. In particular, higher energy efficiency may only be achieved through the purchase of new equipment with higher capital costs than those of less efficient models. Hence, estimates of the direct rebound effect that rely primarily upon historical and/or cross-sectional variations in energy prices could overestimate the direct rebound effect because the additional capital costs required to improve energy efficiency will not be taken into account (Henly, Ruderman, and Levine, 1988).

Second, energy price elasticities tend to be higher for periods with rising prices than for those with falling prices (Dargay and Gately 1994, 1995; Gately 1992, 1993; Haas and Schipper 1998). For example, Dargay (1992) found that the reduction in UK energy demand following the price rises of the late 1970s was five times greater than the increase in demand following the price collapse of the mid-1980s. An explanation may be that higher energy prices induce technological improvements in energy efficiency, which may also become embodied in regulations (Grubb 1995). Also, investment in measures such as thermal insulation is largely irreversible over the short- to medium-term. But the appropriate proxy for improvements in energy efficiency is *reductions* in energy prices. As many studies based upon time-series data incorporate periods of rising energy prices, the estimated price elasticities may overestimate the response to falling energy prices. As a result, such studies could overestimate the direct rebound effect.

Third, while improved energy efficiency may increase the demand for useful work (e.g., you could drive farther after purchasing an energy-efficient car), it is also possible that the anticipated high demand for useful work may increase the demand for energy efficiency (e.g., you purchase an energy-efficient car because you expect to drive farther). In these circumstances, the demand for useful work depends on the energy cost of useful work, which depends upon energy efficiency, which depends upon the demand for useful work (Small and Van Dender 2005). Hence, the direct rebound effect would not be the only explanation for any measured correlation between energy efficiency and the demand for useful work. This so-called endogeneity can be addressed through the use of simultaneous equation models, but these are relatively uncommon owing to their greater data requirements. If, instead, studies include the endogenous variable(s) within a single equation and do not use appropriate techniques to estimate this equation, the resulting estimates could be biased. Several studies of direct rebound effects could be flawed for this reason.

Table 13.1. Estimates of the Long-Run Direct Rebound Effect for Consumer Energy Services in OECD Countries

End-use	Range of values in evidence base (%)	'Best guess' (%)	No. of studies	Degree of confidence
Personal automotive transport	3–87	10–30	17	High
Space heating	0.6–60	10–30	9	Medium
Space cooling	1–26	1–26	2	Low
Other consumer energy services	0–41	<20	3	Low

be expected to decline in the future as demand saturates and income increases. Both theoretical considerations and the limited empirical evidence suggest that direct rebound effects are significantly smaller for other consumer energy services. However, the same conclusion may not follow for energy efficiency improvements by producers or for low income households in developing countries. Moreover, the evidence base is sparse and has a number of important limitations, including the neglect of marginal consumers, the relatively limited time periods over which the effects have been studied, and the restricted definitions of useful work that have been employed. For these and other reasons, it would be inappropriate to draw conclusions about rebound effects as a whole from this evidence.

Evidence for Indirect and Economy-Wide Rebound Effects

This section summarizes the results of a limited number of studies that provide quantitative estimates of indirect and economy-wide rebound effects.

Indirect rebound effects derive from two sources: the energy required to produce and install the measures that improve energy efficiency, such as thermal insulation; and the indirect energy consumption that results from such improvements. The first of these (*embodied energy*) relates to energy consumption that occurs prior to the energy efficiency improvement, while the second (*secondary effects*) relates to energy consumption that follows the improvement.

Many improvements in energy efficiency can be understood as the substitution of capital for energy within a particular system boundary. For example, thermal insulation (capital) may be substituted for fuel to maintain the internal temperature of a building at a particular level. These possibilities form the basis of estimated energy-saving potentials in different sectors. However, estimates of energy savings typically neglect the energy consumption that is required to produce and maintain the relevant capital – frequently referred to as *embodied energy*. For example, energy is required to produce and install home insulation materials and energy efficient motors. Substituting capital for energy therefore shifts energy use from the sector in which it is used to sectors of the economy that produce that capital. As a result, energy use may increase elsewhere in the economy (Kaufmann and Azary-Lee 1990).

In contrast to other sources of the economy-wide rebound effect, the contribution from embodied energy may be expected to be smaller in the long term than in the short term. This is because the embodied energy associated with capital equipment is analogous to a capital cost and hence diminishes in importance relative to ongoing energy savings as the lifetime of the investment increases. An assessment of the embodied energy associated with a particular energy efficiency improvement should also take into account the relevant alternatives. For example, a mandatory requirement to replace existing refrigerators with more energy efficient models may either increase or decrease aggregate energy consumption over a particular period of time, depending upon the age of the existing stock, the lifetime of the new stock, and the direct and indirect energy consumption associated with different models of refrigerator. In practice, however, such estimates are rare.

Unlike embodied energy, *secondary effects* follow the energy efficiency improvement and result from the induced changes in demand for other goods and services. For example, the diffusion of more fuel-efficient cars may reduce demand for public transport but at the same time increase demand for leisure activities that can only be accessed with a private car. Each of these goods and services will have an indirect energy consumption associated with them, and the changed pattern of demand may act to either increase overall energy consumption or reduce it.

Very similar effects will result from energy efficiency improvements by producers. For example, energy efficiency improvements in steel production should reduce the cost of steel and (assuming these cost reductions are passed on in lower product prices) reduce the input costs of manufacturers that use steel. This in turn should reduce the cost of steel products and increase demand for those products. Such improvements could, for example, lower the cost of passenger cars, increase the demand for car travel, and thereby increase demand for motor fuel.

This example demonstrates how energy efficiency improvements could lead to a series of adjustments in the prices and quantities of goods and services supplied throughout an economy. If the energy efficiency improvements are widespread, the price of energy intensive goods and services may fall to a greater extent than that of non-energy intensive goods and services, thereby encouraging consumer demand to shift towards the former. If energy demand is reduced, the resulting fall in energy prices will encourage greater energy consumption by producers and consumers and will feed through into lower product prices, thereby encouraging further shifts towards energy intensive commodities. Reductions in both energy prices and product prices will increase consumers' real income, thereby increasing demand for products, encouraging investment, stimulating economic growth, and further stimulating the demand for energy. In some circumstances, such improvements could also change trade patterns and international energy prices and therefore have an impact on energy consumption in other countries.

A number of analysts have claimed that the secondary effects from energy efficiency improvements in consumer technologies are relatively small (Greening and Greene 1998; Lovins, Henly, Ruderman, and Levine 1988; Schipper and Grubb 2000). This is because energy makes up a small share of total consumer expenditure, and the energy content of most other goods and services is also small.[11] Analogous arguments apply to the secondary effects for producers: since energy forms a small share of total production costs for most firms and sectors (typically < 3%), and since intermediate goods form a small share of the total costs of most final products, the product of these suggests an indirect effect that is much smaller than the direct effect (Greening and Greene 1998). However, while plausible, these arguments are not supported by the results of several of the energy modelling studies reviewed below. In addition, they assume that the only effect of the energy efficiency improvement is to reduce expenditure on energy. But improvements in the energy efficiency of production processes are frequently associated with improvements in the productivity of capital and labour as well and therefore lead to cost savings that exceed the savings in energy costs alone. In some cases, similar arguments may apply to energy efficiency improvements by consumers; for example, a shift from car travel to cycling could save on depreciation and maintenance costs for vehicles as well as motor fuel costs (Alfredsson 2004). In these circumstances, the secondary effects that result from the adoption of a particular technology could be substantial.

Estimates of Indirect and Economy-Wide Rebound Effects

There are two broad approaches to estimating indirect and economy-wide rebound effects, namely embodied energy estimates and energy modelling.

Embodied Energy Estimates

Some indication of the importance of embodied energy may be obtained from estimates of the own-price elasticity of *aggregate* primary, secondary, or final energy demand in a national economy. In principle, this measures the scope for substituting capital, labour, and materials for energy, while holding output constant. Most energy price elasticities are estimated at the level of individual sectors and therefore do not reflect all the embodied energy associated with capital, labour, and materials inputs. Since the own-price elasticity of aggregate energy reflects this indirect energy consumption, it should in principle be smaller than a weighted average of energy demand elasticities within each sector. However, the aggregate elasticity may also reflect price-induced changes in economic structure and product mix, which in principle could make it larger than the average of sectoral elasticities (Sweeney 1984). These two mechanisms could therefore act in opposition.

Based in part upon modelling studies, Sweeney (1984) puts the long-run elasticity of demand for primary energy in the range of −0.25 to −0.6. In contrast, Kaufmann (1992) uses econometric analysis to propose a range from −0.05 to −0.39, while Hong (1983) estimates a value of −0.05 for the U.S. economy. A low value for this elasticity may indicate a limited scope for substitution and hence the potential for large indirect rebound effects.[12] However, this interpretation is not straightforward, because both direct rebound effects and changes in trade patterns may contribute to the behaviour being measured. Also, measures of the quantity and price of aggregate energy are sensitive to the methods chosen for aggregating the prices and quantities of individual energy carriers, while the price elasticity will also depend upon the particular composition of price changes (for example, increases in oil prices relative to gas) (EMF 4 Working Group 1981). In particular, when different energy types are weighted by their relative marginal productivity, the estimated elasticities tend to be lower (Hong 1983). As a result, the available estimates of aggregate price elasticities may be insufficiently precise to provide much indication of the magnitude of indirect rebound effects.

Relatively few empirical studies have estimated the embodied energy associated with specific energy efficiency improvements, and those that have appear to focus disproportionately upon domestic buildings. In a rare study of energy efficiency improvements by producers, Kaufmann and Azary-Lee (1990) estimates that in the U.S. forest products industry over the period from 1954 to 1984, the embodied energy associated with capital equipment offset the direct energy savings from that equipment by as much as 83% (box 13.7). However, since their methodology is crude and the results specific to the U.S. context, this study provides little indication of the magnitude of these effects more generally.

Estimates of the embodied energy of different categories of goods and services can be obtained from input-output analysis, life-cycle analysis (LCA), or a combination of the two (Chapman 1974; Herendeen and Tanak 1976; Kok, Benders, and Moll 2006). A full life-cycle analysis is time consuming to conduct and must address problems of truncation (that is, uncertainty over the appropriate system boundary)[13] and joint production (that is, how to attribute energy consumption to two or more products from a single sector) (Leach 1975; Lenzen and Dey 2000). Hence, many studies combine standard economic input-output tables with additional information on the energy consumption of individual sectors in order to give a comprehensive and reasonably accurate representation of the direct and indirect energy required to produce rather

Box 13.7. Limits to Substitution for Producers

Kaufmann and Azary-Lee (1990) examined the embodied energy associated with energy efficiency improvements in the U.S. forest products industry over the period from 1954 to 1984. First, they estimated a production function for the output of this industry and used this to derive the 'marginal rate of technical substitution' (MRTS) between capital and energy in a given year – in other words, the amount of gross fixed capital that was used to substitute for a thermal unit of energy in that year. Second, they approximated the embodied energy associated with that capital by means of the ratio of aggregate energy to GDP for the U.S. economy in that year – hence ignoring the particular type of capital used as well as the difference between the energy intensity of the capital-producing sectors and that of the economy as a whole. The product of these two variables gave an estimate of the indirect energy consumption associated with the gross capital stock used to substitute for a unit of energy. This was then multiplied by a depreciation rate to give the energy associated with the capital services used to substitute for a unit of energy.

Finally, they compared the estimated indirect energy consumption with the direct energy savings in the forest products sector in each year. Their results showed that the indirect energy consumption of capital offset the direct savings by between 18% and 83% over the period in question, with the net energy savings generally decreasing over time. The primary source of the variation was the increase in the MRTS over time, implying that an increasing amount of capital was being used to substitute for a unit of energy. However, the results were also influenced by the high ratio of energy to GDP of the U.S. economy, which is approximately twice that of many European countries. Overall, the calculations suggest that the substitution reduced aggregate U.S. energy consumption, but by much less than a sector-based analysis would suggest. Also, their approach did not take into account any secondary effects resulting from the energy efficiency improvements.

The simplicity of this approach suggests the scope for further development and wider application. Accuracy could be considerably improved by the use of more flexible production functions and more precise estimates for the indirect energy consumption associated with specific types and vintages of capital goods. However, to date no other authors appear to have applied this approach to particular industrial sectors or to have related it to the broader debate on the rebound effect.

aggregated categories of goods and services. More detailed, LCA-based estimates are available for individual products such as building materials, but results vary widely from one context to another depending upon factors such as the fuel mix for primary energy supply (Sartori and Hestnes 2007).

As an illustration, Sartori and Hestnes (2007) reviews sixty case studies of buildings and finds that the share of embodied energy in life-cycle energy consumption ranged between 9% and 46% for low energy buildings and between 2% and 38% for conventional buildings, with the wide range reflecting different building types, material choices, and climatic conditions. Two studies that controlled for these variables found that low energy designs could achieve substantial reductions in operating energy consumption with relatively small increases in embodied energy, leading to payback periods for energy saving of as little as one year (Feist 1996; UK Royal Commission on Environmental Pollution 2007; Winther and Hestnes 1999). However, Casals (2006) shows how the embodied energy of such buildings could offset operational energy savings, even with an assumed one-hundred-year lifetime. Such calculations typically neglect differences in energy quality, and the results are sensitive to context, design, and building type. Moreover, similar estimates have not been developed in a systematic fashion for other types of energy efficiency improvement.

By combining estimates of the embodied energy associated with different categories of goods and services with survey data on household consumption patterns, it is possible to estimate the total (direct plus indirect) energy consumption of different types of household, together with the indirect energy consumption associated with particular categories of expenditure (Kok, Benders, and Moll 2006). If this data were available at a sufficiently disaggregated level, it could also be used to estimate the secondary effects associated with energy efficiency improvements by households – provided that additional information is available on either the cross-price elasticity between different product and service categories or the marginal propensity to spend[14] of different income groups. By combining the estimates of embodied energy and secondary effects, an estimate of the total indirect rebound effect may be obtained. Such approaches are static in that they do not capture the full range of price and quantity adjustments, but could nevertheless be informative. However, of the nineteen studies in this area reviewed in Kok, Benders, and Moll (2006), only three were considered to have sufficient detail to allow the investigation of such micro-level changes.

A rare example of this approach is Brännlund, Ghalwash, and Nordstrom (2007), which examines the effect of a 20% improvement in the energy efficiency of personal transport (all modes) and space heating in Sweden. The study estimates an econometric model of aggregate household expenditure, in which the share of total expenditure for thirteen types of good or service is expressed as a function of the total budget, the price of each good or service, and an overall price index. This allows the own-price, cross-price, and income elasticities of each good or service to be estimated. By combining estimated changes in demand patterns with CO_2 emission coefficients for each category of good and service (based upon estimates of direct and indirect energy consumption), Brännlund et al. estimates that energy efficiency improvements in transport and heating lead to economy-wide rebound effects (in carbon terms) of 120% and 170% respectively. These results contradict the econometric evidence on direct rebound effects reviewed above because carbon emissions for heating and transport are estimated to increase. The study also lacks transparency and employs an iterative estimation procedure that is truncated at the first estimation step. A comparable study in Mizobuchi (2007) overcomes some of these weaknesses but nevertheless estimates broadly comparable rebound effects for Japanese households.

In summary, while techniques based upon embodied energy estimates provide a promising approach to quantifying indirect and economy-wide rebound effects, the application of these approaches remains in its infancy.

Energy Modelling Estimates. Embodied energy estimates are less useful for quantifying rebound effects from energy efficiency improvements by producers. In this case, a more suitable approach is to use energy-economic models of the macroeconomy (Grepperud and Rasmussen 2004). Such models are widely used within energy studies (Bhattacharyya 1996) but have only recently been applied to estimate rebound effects. The literature is therefore extremely sparse but now includes two insightful studies commissioned by the UK government (Allan et al. 2006; Barker and Foxon 2006). A key distinction is made between computable general equilibrium (CGE) models of the macroeconomy and models based upon econometrics.

Computable general equilibrium models are widely used in energy studies, partly as a consequence of the ready availability of modelling frameworks and the associated benchmark data. This approach is informed by neoclassical economic theory but can deal with circumstances that are too complex for analytical solutions. The CGE models are calibrated to reflect the structural and behavioural characteristics of particular economies and in principle can indicate the approximate order of magnitude of direct and indirect rebound effects from specific energy efficiency improvements. A CGE model should allow the impacts of such improvements to be isolated, because the counterfactual is simply a model run without any changes in energy efficiency, as well as allowing the rebound effect to be decomposed into its constituent components, such as substitution and output effects (box 13.1). In principle, CGE models also provide scope for sensitivity analysis, although in practice this appears to be rare.

The CGE models have a number of important limitations that have led many authors to question their realism and policy relevance (Barker 2005). While developments in CGE methodology are beginning to overcome some of these weaknesses, most remain. Also, the predictive power of such models is rarely tested, and different models appear to produce widely varying results for similar policy questions (Conrad 1999). Hence, while CGE models may provide valuable insights, the quantitative results of such models should be interpreted with caution.

Allan et al. (2007) identifies and reviews eight CGE modelling studies of economy-wide rebound effects (table 13.2). The models vary considerably in terms of the production functions used, the manner in which different inputs are combined (the 'nesting' structure), the assumed scope for substitution between different inputs, the treatment of labour supply, the manner in which government savings are recycled, and other key parameters. All the studies simulate energy efficiency improvements as energy-augmenting technical change, but some introduce an improvement across the board while others introduce a specific improvement in an individual sector, or combination of sectors. This diversity, combined with the limited number of studies available, makes it difficult to draw any general conclusions.

Table 13.2. Summary of CGE Studies of Rebound Effects

Author/Date	Country or region	Nesting structure	ESUB with energy	Assumed energy efficiency improvements
Semboja (1994)	Kenya	Elec, fuel, K, L	1.0	Two scenarios: an improvement of energy production efficiency and an improvement in energy use efficiency
Dufournaud et al. (1994)	Sudan	Utility functions	0.2 and 0.4	100%, 150%, and 200% improvement in efficiency of wood-burning stoves
Vikstrom (2005)	Sweden	(KE)L	Values range from 0.07 to 0.87	15% increase in energy efficiency in non-energy sectors and 12% increase in energy sectors
Washida (2004)	Japan	(KL)E	0.5	1% in all sectors modelled as change in efficiency factor for use of energy in production
Grepperud and Rasmussen (2004)	Norway	(KE)L	Between 0 and 1	Doubling of growth rates of energy productivity. Four sectors have electricity efficiency doubled, and two have oil efficiency doubled
Glomsrød and Wei (2005)	China	Elec, fuel, K, L	1.0	Business-as-usual scenario compared to case where costless investments generate increased investments and productivity in coal cleaning – lowering price and increasing supply
Hanley et al. (2005)	Scotland	(KL)(EM)	0.3	5% improvement in efficiency of energy use across all production sectors
Allan et al. (2006)	UK	(KL)(EM)	0.3	5% improvement in efficiency of energy use across all production sectors (including energy sectors).

Notes: The production functions combine inputs into pairs, or 'nests.' For example, a nested production (*KL*)*E*; (*KE*)*L*. ESUB means the elasticity of substitution between energy and other inputs. Interpretation between *L* and *E*, while for (*KL*)*E*, it refers to elasticity substitution between (*KL*) and *E*.

The most notable result is that all of the studies find economy-wide rebound effects to be greater than 37%, and most studies show either large rebounds (> 50%) or backfire. The latter was found in two studies of economies in which energy forms an important export and import commodity, suggesting that this is a potentially important and hitherto neglected variable. Allan et al. (2006) finds a long-term rebound effect of 37% from across-the-board improvements in the energy efficiency of UK production sectors, including primary energy supply. This study is summarized in box 13.8.

All but one of the models explore the implications of energy efficiency improvements in production sectors, and the CGE literature offers relatively little insight into the implications of energy efficiency im-

Estimated rebound effect	Comments
>100%	Intuitive presentation, no sensitivity tests, lack of transparency
47%–77%	Models efficiency improvements in domestic stoves. Wide range of sensitivity tests, and good explanation of the factors at work
50%–60%	Applies to 1957–1962 period in which known changes in energy efficiency productivity and structure are combined in turn. Results apply only to energy efficiency component
53% in central scenario	Presentation unclear, although there is some sensitivity analysis, including varying elasticity of substitution from 0.3 to 0.7 jointly with other parameters. Rebound effect increases as energy/value added, labour/capital, and level of energy composite substitution elasticities increase
Small for oil but >100% for electricity	Model is simulated dynamically with a counterfactual case in which projections of world economic growth, labour force growth, technological progress, and net foreign debt are assumed until 2050
>100%	Coal intensive sectors benefit, as does whole economy due to high use of coal in primary energy consumption. Also examines cases where coal use is subject to emissions tax
>100%	Region is significant electricity exporter, and result depends in part on increased electricity exports
37% in central scenario	See box 13.8

function with capital (*K*), labour (*L*) and energy (*E*) inputs could take one of three forms: namely, *K*(*LE*);
depends upon the nesting structure. For example, for *K*(*LE*), ESUB refers to the elasticity of substitution

provements in consumer goods. As there are differences across income groups, this would require a much greater detail on the demand side of the CGE models than is commonly the case. Also, most of these studies assume that energy efficiency improvements are costless. Only Allan et al. (2006) considers the implications of additional costs associated with energy efficiency improvements, and it finds that rebound effects are correspondingly reduced.

One final approach to estimating economy-wide rebound effects is through the use of macroeconometric models of national economies. These can overcome several of the weaknesses of CGE modelling while at the same time providing a greater level of disaggregation that permits the investigation of specific government policies. In contrast to

Box 13.8. CGE Estimates of Economy-Wide Rebound Effects for the United Kingdom

Allan et al. (2006) estimates economy-wide rebound effects for the United Kingdom, following a 5% across-the-board improvement in the efficiency of energy use in all production sectors. As the model allows for the gradual updating of capital stocks, the study is able to estimate a short-run rebound effect of 50% and a long-run effect of 37%.

The energy efficiency improvements increase long-run GDP by 0.17% and employment by 0.21%. They have a proportionally greater impact on the competitiveness of energy intensive sectors, which is passed through in lower product prices despite a 0.3% increase in real wages. Output is increased in all sectors, with the iron and steel and the pulp and paper sectors benefiting the most with long-run increases of 0.67% and 0.46%, respectively. In contrast, the output of the oil refining and electricity industries (i.e., oil and electricity demand) is reduced, with the price of conventional electricity falling by 24% in the long run. This fall in energy prices contributes a significant proportion of the overall rebound effect and results from both cost reductions in energy production – due to the energy efficiency improvements – and reduced energy demand.

In practice, a 5% improvement in energy efficiency may not be feasible for an industry such as electricity generation which is operating close to thermodynamic limits. It would also require major new investment and take time to be achieved. Allan et al.'s results suggest that the rebound effect would be smaller if energy efficiency improvements were confined to energy users, but the importance of this cannot be quantified. Moreover, the results demonstrate that energy efficiency improvements in the energy supply industry may be associated with large rebound effects.

A notable feature of this study is the use of sensitivity tests. Varying the assumed elasticity of substitution between energy and non-energy inputs between 0.1 and 0.7 (compared to a baseline value of 0.3) had only a small impact on economic output (from 0.16% in the low elasticity case to 0.10% in the high case) but had a major impact on the rebound effect. This varied from

7% in the low case to 60% in the high case. Grepperud and Rasmussen (2004) reports similar results, which highlights the importance of this parameter for CGE simulations. Unfortunately, the empirical basis for the assumed parameter values in CGE models is extremely weak, while the common assumption that such parameters are constant is flawed (Broadstock, Hunt, and Sorrell 2007).

Varying the elasticity of demand for exports was found to have only a small impact on GDP and energy demand, suggesting that the energy efficiency improvements had only a small impact on the international competitiveness of the relevant industries. However, different treatments of the additional tax revenue were found to be important.

their CGE counterparts, macroeconometric models do not rely upon restrictive assumptions such as constant returns to scale and perfect competition, and replace the somewhat ad hoc use of parameter estimates with econometric equations estimated for individual sectors. However, this greater realism is achieved at the expense of greater complexity and more onerous data requirements.

At the time of writing, Barker and Foxon (2006) provides the only example of the application of such models to economy-wide rebound effects. The MDM-E3 model was used to simulate the macroeconomic impact of a number of UK energy efficiency policies over the period from 2000 to 2010 The study combined exogenous estimates of 'gross' energy savings and direct rebound effects with modelling of indirect effects. The direct rebound effects were estimated to reduce overall energy savings by 15%, while the indirect effects reduced savings by a further 11% – leading to an estimated economy-wide rebound effect of 26% in 2010. The indirect effects were higher in the energy intensive industries (25%) and lower for households and transport (7%). The primary source of the indirect effects was substitution between energy and other goods by households, together with increases in output by (particularly energy intensive) industry, which in turn led to increased demand for both energy and energy intensive intermediate goods. Increases in consumers' real income contributed relatively small rebound effects (0.2%).

However, there are a number of reasons that this study may have underestimated the economy-wide effects. First, while output effects were estimated, the substitution between (cheaper) energy services and other inputs was ignored. Second, the modelling implicitly assumed 'pure' energy efficiency improvements, with no associated improvements in the productivity of other inputs. But if energy efficient technologies are commonly associated with such improvements, rebound effects could be larger. Third, the model did not reflect the indirect energy consumption embodied within the energy efficient technologies themselves. Finally, the study confined attention to national energy use and ignored the indirect energy consumption associated with increased imports and tourism. This omission could be significant from a climate change perspective because this corresponds to approximately ~40% of the extra domestic output.[15]

In summary, given the small number of studies available, the diversity of approaches, and the methodological weaknesses associated with each, it is not possible to draw any general conclusions regarding the size of the economy-wide rebound effect from either embodied energy or energy modelling studies. Indeed, the most important insight is that the economy-wide rebound effect varies greatly from one circumstance to another; therefore, general statements on the size of such effects are misleading. It is notable, however, that the available studies suggest that economy-wide effects are frequently large (that is, > 50%) and that the potential for backfire cannot be ruled out. Moreover, these estimates derive from pure energy efficiency improvements and therefore do not rely upon simultaneous improvements in the productivity of capital and labour inputs.

Policy Implications of Rebound Effects

Energy efficiency may be encouraged through policies that raise energy prices, such as carbon taxes, or through non-price policies such as building regulations. Appropriately designed, such policies should continue to play an important role in energy and climate policy. However, the neglect of rebound effects has meant that many official and independent appraisals have overstated the potential contribution of non-price policies to reducing energy consumption and carbon emissions. Taking rebound effects into account will reduce the apparent effectiveness of such policies.

At the same time, many energy efficiency opportunities appear to be highly cost-effective for the individuals and organizations involved (IPCC 2007). If non-price policies can cost-effectively overcome the various market and organizational failures that prevent such opportunities from being taken up, they should increase the real income and welfare of consumers and improve productivity in both the public and the private sectors. If economy-wide rebound effects are less than unity, they may also be cost-effective from a climate change perspective – for example, on the basis of the marginal abatement cost of carbon emissions. They may not, however, reduce energy consumption and carbon emissions by as much as has been previously assumed.

While rebound effects cannot be quantified with much confidence, there should be scope for including estimated effects within policy appraisals and for building headroom into policy targets to allow for such effects. There may also be scope for using these estimates to target policies more effectively – for example, by prioritizing sectors and technologies where rebound effects are expected to be smaller. In contrast, where rebound effects are expected to be large, there may be a greater need for policies that increase energy prices.

A particularly important finding from the UK Energy Research Centre's review is that technologies that reduce capital and labour costs as well as energy costs may be associated with the largest rebound effects. This includes the win-win opportunities that are regularly highlighted by energy efficiency advocates. For example, Lovins and Lovins (1997) used case studies to argue that better visual, acoustic, and thermal comfort in well-designed, energy efficient buildings can improve labour productivity by as much as 16%. Since labour costs in commercial buildings are typically twenty-five times greater than energy costs, the resulting cost savings can dwarf those from reduced energy consumption. However, if the total cost savings are twenty-five times greater, the indirect rebound effect may be twenty-five times greater as well. While identifying such situations is far from straightforward, the implications of encouraging such opportunities should ideally be taken into account. It may make more sense to focus non-price policy on dedicated energy efficient technologies, such as thermal insulation, that do not have these wider benefits.

The primary source of rebound effects is the reduced cost of energy services. In principle, policies such as carbon taxes should reduce direct and indirect rebound effects by ensuring that the cost of energy services remains relatively constant while energy efficiency improves. But carbon pricing will need to increase over time at a rate sufficient to accommodate both income growth and rebound effects, simply to prevent carbon emissions from increasing. It will need to increase more rapidly if emissions are to be reduced (Birol and Keppler 2000).

Perhaps the most attractive approach to mitigating rebound effects is the use of cap-and-trade schemes for carbon emissions. The impact of these may helpfully be illustrated with the use of the so-called IPAT equation (Chertow 2001). For example, economy-wide carbon emissions (I) may be expressed as the product of population (P), GDP per capita ($A = Y/P$), and carbon emissions per unit of GDP ($T = C/Y$): $I = PAT$. Non-price energy efficiency policies seek to reduce carbon intensity (T) and thereby carbon emissions (I). But there may be feedback effects between the right-hand side variables in the form of various rebound effects (Alcott 2008). In particular, energy efficiency may encourage economic growth (A), which in turn will increase the total demand for energy and hence carbon emissions (I). Improvements in energy efficiency could therefore have unintended consequences that undermine the policy objective. In contrast, by placing a quantitative limit upon carbon emissions, a cap-and-trade scheme focuses directly on the relevant environmental impact (I) rather than on one of the factors contributing to that impact. Provided the scheme is effectively enforced, it should provide a guarantee that the desired environmental outcome is achieved (although there may be unintended consequences for sources of emissions that are not covered by the scheme). A cap-and-trade scheme therefore provides some insurance against rebound effects. The downside, of course, is that a cap on carbon emissions (I) could have unintended consequences for the right-hand side variables, notably per capita GDP (A). The importance of this is greatly disputed but will depend in part upon how the scheme is implemented – including in particular the choice between free allocation of allowances and revenue neutral auctioning (Bovenberg 1999).

Policies such as carbon taxes and cap-and-trade schemes may be insufficient on their own because they will not overcome the numerous barriers to the innovation and diffusion of low carbon technologies and could in some circumstances have adverse impacts on income distribution and competitiveness (Sorrell and Sijm 2003). Similarly, non-price policies to address market barriers may also be insufficient because rebound effects could offset some or all of the energy savings. Effective climate policy will therefore need to combine both.

Conclusions

This chapter has clarified the definition of direct, indirect, and economy-wide rebound effects, highlighted the methodological challenges associated with quantifying such effects, summarized the estimates that are currently available, and highlighted some relevant policy implications. The main conclusions are as follows:

1. *Rebound effects are significant, but they need not make energy efficiency policies ineffective in reducing energy demand.* Rebound effects vary widely between different technologies, sectors, and income groups and in most cases cannot be quantified with much confidence. However, the evidence does not suggest that improvements in energy efficiency routinely lead to economy-wide increases in energy consumption, as some commentators have suggested. At the same time, the evidence does not suggest that rebound effects are small (for example, < 10%) as many analysts and policy makers assume.
2. *For most consumer energy services in OECD countries, direct rebound effects are likely to exceed 30%.* Improvements in energy efficiency should achieve 70% or more of the expected reduction in energy consumption for those services, although the existence of indirect effects means that the economy-wide reduction in energy consumption will be less. However, these conclusions cannot be extended to producers or to households in developing countries. These conclusions are also subject to a number of important qualifications, including the neglect of marginal consumers and the relatively limited time period over which the effects have been studied.
3. *There are relatively few quantitative estimates of indirect and economy-wide rebound effects, but several studies suggest that the economy-wide effect may often exceed 50%.* The magnitude of economy-wide effects depends very much upon the sector where the energy efficiency improvement takes place and is sensitive to a number of variables. A handful of modelling studies estimate economy-wide rebound effects of 26% or more, with half of the studies predicting backfire. These effects derive from pure energy efficiency improvements by producers (not consumers) and therefore do not rely upon simultaneous improvements in the productivity of other inputs. However, the small number of studies available, the diversity of approaches used, and the variety of methodological weaknesses associated with the CGE approach all suggest the need for caution when interpreting these results.
4. *If rebound effects are neglected, energy efficiency policies will fail.* Energy efficiency may be encouraged through policies that raise energy prices, such as carbon taxes, or through non-price policies such as building regulations. Both should continue to play an important

> role in energy and climate policy. However, many official and independent appraisals of such policies have overstated the potential contribution of non-price policies to reducing energy consumption and carbon emissions. While rebound effects cannot be quantified with much confidence, there should be scope for including estimated effects within policy appraisals and using these estimates to target policies more effectively. It may make more sense to focus policy on dedicated energy efficient technologies because win-win technologies that also reduce capital and labour costs may be associated with large rebound effects. In all cases, carbon/energy pricing can mitigate rebound effects by ensuring that the cost of energy services remains relatively constant while energy efficiency improves. Hence, carbon taxes and/or cap-and-trade schemes form an essential component of climate policy at all levels. A policy mix that does not include carbon/energy pricing is liable to fail.

Given the potential importance of rebound effects, the evidence base is remarkably weak. While the precise quantification of rebound effects may be an elusive goal, it should be possible to gain a much better understanding of the determinants of these effects than we have at present – including the conditions under which they are more or less likely to be large. This understanding has been inhibited in the past by confusion over basic definitions, an excessive focus upon theoretical arguments, and an overly polarized debate around the likelihood of backfire. Researchers must move beyond this and conduct more systematic empirical research on this important topic.

Policymakers must also recognize the importance of rebound effects, allow for them within policy appraisals, and take steps to mitigate such effects through carbon/energy pricing. Indeed, given their potential importance for energy and climate policy, the extent to which rebound effects have been neglected is quite remarkable. Something is surely amiss when the most in-depth and influential studies (for example, Stern; Garnaut; and IPCC) overlook this topic altogether. This neglect could stem in part from limited understanding of the issue and the considerable difficulties in quantifying such effects, but it may also result from the challenge that these effects pose to conventional wisdom and of their uncomfortable policy implications. While backfire is not inevitable, it is clearly the case that improvements in energy efficiency do not always provide the low cost, win-win outcomes that are generally assumed. Instead, they have unintended consequences that can frequently undermine the original objectives.

Modern climate policy is predicated on the assumption that it is possible to maintain high levels of economic growth while at the same time making substantial reductions in energy use and carbon emissions. The evidence presented in this chapter does not prove that this is impossible, but it does suggest that it is more difficult to achieve than is generally assumed. Rebound effects therefore have some far-reaching implications. We ignore them at our peril.

NOTES

1 This may be expressed as *REB* = [(*DIR* + *IND*)/*ENG*]* 100%, where *ENG* represents the expected energy savings from a particular energy efficiency improvement without taking rebound effects into account; *DIR* represents the increase in energy consumption resulting from the direct rebound effect; and *IND* represents the increase in energy consumption resulting from the indirect rebound effects.

2 For example, improvements in the energy efficiency of electric motors in the engineering sector may lead to rebound effects within that sector, with the result that the energy intensity of that sector is reduced by less than it would be in the absence of such effects. But if the energy intensity of the sector is taken as the independent variable, these lower-level rebound effects will be overlooked.

3 A common measure of the ability to perform useful work is *exergy*, defined as the maximum amount of work obtainable from a system as it comes (reversibly) to equilibrium with a reference environment (Wall 2004). Exergy is only non-zero when the system under consideration is distinguishable from its environment through differences in relative motion, gravitational potential, electromagnetic potential, pressure, temperature, or chemical composition. Unlike energy, exergy is 'consumed' in conversion processes and is mostly lost in the form of low temperature heat. A heat unit of electricity, for example, will be ranked higher on an exergy basis than will a heat unit of oil or natural gas because a heat unit of electricity can do more useful work.

4 The marginal product of an energy input into a production process is the marginal increase in the value of output produced by the use of one additional heat unit of energy input. In the absence of significant market distortions, the relative price per kilowatt-hour of different energy carriers can provide a broad indication of their relative marginal productivities (Kaufmann 1994).

5 For example, on a thermal input basis, per capita energy consumption in

the U.S. residential sector decreased by 20% over the period from 1970 to 1991, but when adjustments are made for changes in energy quality (notably the increasing use of electricity), per capita energy consumption is found to have increased by 7% (Zarnikau, Guermouches, and Schmidt 1996). This difference demonstrates that technical progress in energy use is not confined to improvements in thermodynamic efficiency but also includes the substitution of low quality fuels by high quality fuels (notably electricity), thereby increasing the amount of utility or economic output obtained from the same heat content of input (Kaufmann 1992).

6 If energy efficiency were measured instead as tonne-kilometres per litre of fuel, rebound effects would show up as an increase in tonne-kilometres driven, which may be decomposed into the product of the number of vehicles, the mean vehicle weight, and the mean distance travelled per vehicle per year. To the extent that vehicle weight provides a proxy for factors such as comfort, safety, and carrying capacity, this approach effectively incorporates some features normally classified as attributes of the energy service into the measure of useful work. It also moves closer to a thermodynamic measure of energy efficiency by focusing upon the movement of mass rather than the movement of people.

7 For example, if energy accounts for 50% of the total cost of an energy service, doubling energy efficiency will reduce the total costs of the energy service by 25%. But if energy only accounts for 10% of total costs, doubling energy efficiency will reduce total cost by only 5%. In practice, improvements in energy efficiency may themselves be costly.

8 For example, daily average household temperatures will generally increase following improvements in thermal insulation, even if the heating controls remain unchanged. This is because insulation contributes to a more even distribution of warmth around the house, reduces the rate at which a house cools down when the heating is off, and delays the time at which it needs to be switched back on (Milne and Boardman 2000).

9 The rationale for the use of these elasticities, and the relationship between them, is explained in detail in Sorrell and Dimitropoulos (2007a).

10 Under these circumstances: $\eta_\varepsilon(E) = -1$. A positive rebound effect implies that $\eta_\varepsilon(S) > 0$ and $0 > \eta_\varepsilon(S) > -1$, while backfire implies that $\eta_\varepsilon(S) > 1$ and $\eta_\varepsilon(E) > 0$.

11 For example, suppose energy efficiency improvements reduce natural gas consumption per unit of space heated by 10%. If there is no direct rebound effect, consumers will reduce expenditure on natural gas for space heating by 10%. If natural gas for heating accounts for 5% of total consumer expenditure, consumers will experience a 0.5% increase in their real dispos-

able income. If *all* of this were spent on motor fuel for additional car travel, the net energy savings (in kWh thermal content) would depend upon the ratio of natural gas prices to motor fuel prices and could in principle be more or less than one. In practice, however, motor fuel only accounts for a portion of the total cost of car travel, and car travel only accounts for a portion of total consumer expenditure. For the great majority of goods and services, input-output data suggest that the effective expenditure on energy should be less than 15% of the total expenditure. Hence, by this logic, the secondary effect should be only around one-tenth of the direct effect (Greening and Greene 1998).

12 This is in contrast to the own-price elasticity of energy demand for an individual energy service, where high values may indicate the potential for large direct rebound effects.

13 For example, should the indirect energy costs of a building also include the energy used to make the structural steel and mine the iron ore used to make the girders? This is referred to as the truncation problem because there is no standard procedure for determining when energy costs become small enough to neglect.

14 The marginal propensity to spend is defined as the change in expenditure on a particular product or service, divided by the change in total expenditure. The marginal propensity to spend on different goods and services varies with income, and it is an empirical question as to whether the associated indirect energy consumption is larger or smaller at higher levels of income. However, the greater use of energy intensive travel options by high income groups (notably, flying) could be significant in some cases.

15 Also, any similarity between this result and that of Allan et al. (2006) is spurious because they use different approaches to model different types of rebound effect from different types and sizes of energy efficiency improvement in different sectors.

REFERENCES

Alcott, B. 2005. 'Jevons' paradox.' *Ecological Economics* 54 (1): 9–21.

– 2008. 'The sufficiency strategy: Would rich-world frugality lower environmental impact?' *Ecological Economics* 64:770–86.

Alfredsson, E.C. 2004. '"Green" consumption – No solution for climate change.' *Energy* 29:513–24.

Allan, G., M. Gilmartin, P.G. McGregor, K. Swales, and K. Turner. 2007. 'UKERC review of evidence for the rebound effect: Technical Report 4,

Energy-economic modelling studies.' UK Energy Research Centre: London.

Allan, G., N. Hanley, P.G. McGregor, J. Kim Swales, and K. Turner. 2006. 'The macroeconomic rebound effect and the UK economy.' Department of Economics, University of Strathclyde.

Barker, T. 2005. 'The transition to sustainability: A comparison of general equilibrium and space-time economics approaches.' Tyndall Centre for Climate Change Research.

Barker, T., P. Ekins, and N. Johnstone. 1995. *Global warming and energy demand*. London: Routledge.

Barker, T., and T. Foxon. 2006. 'The macroeconomic rebound effect and the UK economy.' 4CMR, University of Cambridge.

Bhattacharyya, S.C. 1996. 'Applied general equilibrium models for energy studies: A survey.' *Energy Economics* 18:145–64.

Birol, F., and J.H. Keppler. 2000. 'Prices, technology development, and the rebound effect.' *Energy Policy* 28 (6-7): 457–69.

Bovenberg, A.L. 1999. 'Green tax reforms and the double dividend: Updated reader's guide.' *International Tax and Public Finance* 6:421–43.

Boyd, G.A., and J.X. Pang. 2000. 'Estimating the linkage between energy efficiency and productivity.' *Energy Policy* 28:289–96.

Brännlund, R., T. Ghalwash, and J. Nordstrom. 2007. 'Increased energy efficiency and the rebound effect: Effects on consumption and emissions.' *Energy Economics* 29 (1): 1–17.

Broadstock, D., L.C. Hunt, and S. Sorrell. 2007. 'UKERC review of evidence for the rebound effect: Technical Report 3, Elasticity of substitution studies.' UK Energy Research Centre, London.

Brookes, L. 2004. 'Energy efficiency fallacies: A postscript.' *Energy Policy* 32 (8): 945–7.

Brookes, L.G. 1984. 'Long-term equilibrium effects of constraints in energy supply.' In *The Economics of Nuclear Energy*, ed. Leonard Brookes and H. Motamen. London: Chapman and Hall.

– 1990. 'The greenhouse effect: The fallacies in the energy efficiency solution.' *Energy Policy* 18 (2): 199–201.

– 2000. 'Energy efficiency fallacies revisited.' *Energy Policy* 28 (6-7): 355–66.

Casals, X.G. 2006. 'Analysis of building energy regulation and certification in Europe: Their role, limitations, and differences.' *Energy and Buildings* 38: 381–92.

Chapman, P. 1974. 'Energy costs: A review of methods.' *Energy Policy* 2 (2): 91–103.

Chertow, M.R. 2001. 'The IPAT equation and its variants: Changing views of technology and environmental impacts.' *Journal of Industrial Ecology* 4 (4): 13–29.

Cleveland, C.J., R.K. Kaufmann, and D.I. Stern. 2000. 'Aggregation and the role of energy in the economy.' *Ecological Economics* 32:301–17.

Conrad, K. 1999. 'Computable general equilibrium models for environmental economics and policy analysis.' In *Handbook of Environmental and Resource Economics*, ed. J.C.J.M van den Bergh. Cheltenham, UK: Edward Elgar.

Dahl, C. 1993. 'A survey of energy demand elasticities in support of the development of the NEMS.' Department of Mineral Economics, Colorado School of Mines, Colorado.

– 1994. 'Demand for transportation fuels: A survey of demand elasticities and their components.' *Journal of Energy Literature* 1 (2): 3.

Dahl, C., and T. Sterner. 1991. 'Analyzing gasoline demand elasticities: A survey.' *Energy Economics* 13 (3): 203–10.

Dargay, J.M. 1992. *Are Price and Income Elasticities of Demand Constant? The UK Experience*. Oxford: Oxford Institute for Energy Studies.

Dargay, J.M., and D. Gately. 1994. 'Oil demand in the industrialised countries.' *Energy Journal* 15 (Special Issue): 39–67.

– 1995. 'The imperfect price irreversibility of non-transportation of all demand in the OECD.' *Energy Economics* 17 (1): 59–71.

Davis, L.W. 2007. 'Durable goods and residential demand for energy and water: Evidence from a field trial.' Department of Economics, University of Michigan.

Dufournaud, C.M., J.T. Quinn, and J.J. Harrington. 1994. 'An applied general equilibrium (AGE) analysis of a policy designed to reduce the household consumption of wood in the Sudan.' *Resource and Energy Economics* 16 (1): 67–90.

EMF 4 Working Group. 1981. 'Aggregate elasticity of energy demand.' *Energy Journal* 2 (2): 37–75.

Espey, J.A., and M. Espey. 2004. 'Turning on the lights: A meta-analysis of residential electricity demand elasticities.' *Journal of Agricultural and Applied Economics* 36 (1): 65–81.

Espey, M. 1998. 'Gasoline demand revisited: An international meta-analysis of elasticities.' *Energy Economics* 20:273–95.

Feist, W. 1996. 'Life-cycle energy balances compared: Low energy house, passive house, self-sufficient house.' In *Proceedings of the International Symposium of CIB W67*. Vienna, Austria.

Frondel, M., and C.M. Schmidt. 2005. 'Evaluating environmental programs: The perspective of modern evaluation research.' *Ecological Economics* 55 (4): 515–26.

Garnaut, Ross. 2008. *The Garnaut climate change review.* Cambridge: Cambridge University Press.

Gately, Dermot. 1992. 'Imperfect price-reversibility of U.S. gasoline demand: Asymmetric responses to price increases and declines.' *The Energy Journal* 13 (4): 179–207.

– 1993. 'The imperfect price reversibility of world oil demand.' *Energy Journal* 14 (4): 163–82.

Glomsrød, S., and T. Y. Wei. 2005. 'Coal cleaning: A viable strategy for reduced carbon emissions and improved environment in China?' *Energy Policy* 33 (4): 525–42.

Graham, D.J., and S. Glaister. 2002. 'The demand for automobile fuel: A survey of elasticities.' *Journal of Transport Economics and Policy* 36 (1): 1–26.

Greene, D.L., J.R. Kahn, and R.C. Gibson. 1999. 'Fuel economy rebound effect for US household vehicles.' *Energy Journal*, 20 (3): 1–31.

Greening, L.A., and D.L. Greene. 1998. 'Energy use, technical efficiency, and the rebound effect: A review of the literature.' Hagler Bailly and Co., Denver.

Greening, L.A., D.L. Greene, and C. Difiglio. 2000. 'Energy efficiency and consumption: The rebound effect; a survey.' *Energy Policy* 28 (6-7): 389–401.

Grepperud, Sverre, and Ingeborg Rasmussen. 2004. 'A general equilibrium assessment of rebound effects.' *Energy Economics* 26 (2): 261–82.

Grubb, M.J. 1995. 'Asymmetrical price elasticities of energy demand.' In *Global warming and energy demand*, ed. T. Barker, P. Ekins, and N. Johnstone. London and New York: Routledge.

Guertin, C., S. Kumbhakar, and A. Duraiappah. 2003. 'Determining demand for energy services: Investigating income-driven behaviours.' International Institute for Sustainable Development.

Haas, R., and L. Schipper. 1998. 'Residential energy demand in OECD-countries and the role of irreversible efficiency improvements.' *Energy Economics* 20 (4): 421–42.

Hanley, M., J.M. Dargay, and P.B. Goodwin. 2002. 'Review of income and price elasticities in the demand for road traffic.' ESRC Transport Studies Unit, University College London.

Hanley, N., P.G. McGregor, J.K. Swales, and K. Turner. 2005. 'Do increases in resource productivity improve environmental quality? Theory and evidence on rebound and backfire effects from an energy-economy-environment regional computable general equilibrium model of Scotland.' Department of Economics, University of Strathclyde.

Harrison, D., G. Leonard, B. Reddy, D. Radov, P. Klevnäs, J. Patchett, and P. Reschke. 2005. 'Reviews of studies evaluating the impacts of motor vehicle greenhouse gas emissions regulations in California.' National Economic Research Associates, Boston.

Hartman, R.S. 1988. 'Self-selection bias in the evaluation of voluntary energy conservation programs.' *The Review of Economics and Statistics* 70 (3): 448–58.

Haughton, J., and S. Sarkar. 1996. 'Gasoline tax as a corrective tax: Estimates for the United States, 1970–1991.' *Energy Journal* 17 (2): 103–26.

Henly, John, Henry Ruderman, and Mark D. Levine. 1988. 'Energy savings resulting from the adoption of more efficient appliances: A follow-up.' *Energy Journal* 9 (2): 163–70.

Herendeen, R., and J. Tanak. 1976. 'The energy cost of living.' *Energy* 1 (2): 165–78.

Herring, Horace. 2006. 'Energy efficiency: A critical view.' *Energy* 31 (1): 10–20.

Hong, N.V. 1983. 'Two measures of aggregate energy production elasticities.' *The Energy Journal* 4 (2): 172–7.

Hunt, L.C., and J. Evans, eds. 2008. *The international handbook of the economics of energy*. Cheltenham, UK: Edward Elgar.

Intergovernmental Panel on Climate Change (IPCC). 2007. 'Climate change 2007: Mitigation of climate change.'

Johansson, O., and L. Schipper. 1997. 'Measuring long-run automobile fuel demand: Separate estimations of vehicle stock, mean fuel intensity, and mean annual driving distance.' *Journal of Transport Economics and Policy* 31 (3): 277–92.

Kaufmann, Robert K. 1992. 'A biophysical analysis of the energy/real GDP ratio: Implications for substitution and technical change.' *Ecological Economics* 6 (1): 35–56.

– 1994. 'The relation between marginal product and price in US energy markets: Implications for climate change policy.' *Energy Economics* 16 (2): 145–58.

Kaufmann, R.K., and I.G. Azary-Lee. 1990. 'A biophysical analysis of substitution: Does substitution save energy in the US forest products industry?' In *Ecological economics: Its implications for forest management and research*, Swedish University of Agricultural Sciences, proceedings of a workshop held in St Paul, Minnesota.

Khazzoom, J.D. 1980. 'Economic implications of mandated efficiency in standards for household appliances.' *Energy Journal* 1 (4): 21–40.

Kok, R., R.M.J. Benders, and H.C. Moll. 2006. 'Measuring the environmental load of household consumption using some methods based on input-output energy analysis: A comparison of methods and a discussion of results.' *Energy Policy* 34 (17): 2744–61.

Leach, G. 1975. 'Net energy analysis: Is it any use?' *Energy Policy* 3 (4): 332–44.

Lenzen, M., and C. Dey. 2000. 'Truncation error in embodied energy analyses of basic iron and steel products.' *Energy* 25 (6): 577–85.
Lovins, A.B. 1998. 'Further comments on Red Herrings.' Letter to the *New Scientist*.
Lovins, Amory B., John Henly, Henry Ruderman, and Mark D. Levine. 1988. 'Energy saving resulting from the adoption of more efficient appliances: Another view; a follow-up.' *The Energy Journal* 9 (2): 155.
Lovins, Amory B., and L.H. Lovins. 1997. 'Climate: Making sense and making money.' Rocky Mountain Institute, Old Snowmass, CO.
Meyer, B. 1995. 'Natural and quasi experiments in economics.' *Journal of Business and Economic Statistics* 13 (2): 151–60.
Milne, Geoffrey, and Brenda Boardman. 2000. 'Making cold homes warmer: The effect of energy efficiency improvements in low-income homes.' *Energy Policy* 28:411–24.
Mizobuchi, K.I. 2007. 'An empirical evidence of the rebound effect considering capital costs.' Graduate School of Economics, Kobe University, Japan.
Nadel, Steven. 1993. 'The take-back effect: Fact or fiction.' American Council for an Energy-Efficient Economy.
Pye, M., and A. McKane. 1998. 'Enhancing shareholder value: Making a more compelling energy efficiency case to industry by quantifying non-energy benefits.' In *Proceedings of 1999 Summer Study on Energy Efficiency in Industry*, 325–36. Washington, DC: American Council for an Energy-Efficient Economy.
Roy, J. 2000. 'The rebound effect: Some empirical evidence from India.' *Energy Policy* 28 (6-7): 433–8.
Sanders, M., and M. Phillipson. 2006. 'Review of differences between measured and theoretical energy savings for insulation measures.' Centre for Research on Indoor Climate and Health, Glasgow Caledonian University.
Sartori, I., and A.G. Hestnes. 2007. 'Energy use in the life-cycle of conventional and low-energy buildings: A review article.' *Energy and Buildings* 39:249–57.
Saunders, Harry D. 1992. 'The Khazzoom-Brookes postulate and neoclassical growth.' *Energy Journal* 13 (4): 131.
– 2000. 'A view from the macro side: Rebound, backfire, and Khazzoom-Brookes.' *Energy Policy* 28 (6-7): 439–49.
– 2007. 'Fuel conserving (and using) production function.' Decision Processes Incorporated, Danville, CA.
Schipper, Lee, and Michael Grubb. 2000. 'On the rebound? Feedback between energy intensities and energy uses in IEA countries.' *Energy Policy* 28 (6-7): 367–88.

Schipper, Lee, M. Josefina, L.P. Figueroa, and M. Espey. 1993. 'Mind the gap: The vicious circle of measuring automobile fuel use.' *Energy Policy* 21 (12): 1173–90.

Semboja, H.H.H. 1994. 'The effects of an increase in energy efficiency on the Kenyan economy.' *Energy Policy* 22 (3): 217–25.

Small, K.A., and K. van Dender. 2005. 'A study to evaluate the effect of reduced greenhouse gas emissions on vehicle miles travelled.' Department of Economics, University of California, Irvine.

Sommerville, M., and S. Sorrell. 2007. 'UKERC review of evidence for the rebound effect: Technical Report 1, Evaluation studies.' UK Energy Research Centre, London.

Sorrell, S. 2007. 'The rebound effect: An assessment of the evidence for economy-wide energy savings from improved energy efficiency.' UK Energy Research Centre, London.

Sorrell, S. and J. Dimitropoulos. 2007a. 'The rebound effect: Microeconomic definitions, limitations and extensions.' *Ecological Economics*, in press.

– 2007b. 'The rebound effect: Microeconomic definitions, limitations and extensions.' *Ecological Economics* 65 (3): 636–49.

– 2007c. 'UKERC review of evidence for the rebound effect: Technical Report 3, Econometric studies.' UK Energy Research Centre, London.

– 2007d. 'UKERC review of evidence for the rebound effect: Technical Report 5, Energy, productivity and economic growth studies.' UK Energy Research Centre, London.

Sorrell, S., and J. Sijm. 2003. 'Carbon trading in the policy mix.' *Oxford Review of Economic Policy* 19 (3): 420–37.

Stern, N. 2007. *Stern Review: The Economics of Climate Change*. HM Treasury: London.

Sweeney, J.L. 1984. 'The response of energy demand to higher prices: What have we learned?' *American Economic Review* 74 (2): 31–7.

United Kingdom. Department of Environment, Food, and Rural Affairs. 2007. 'Consultation document: Energy, cost and carbon savings for the draft EEC 2008–11 illustrative mix.' London.

United Kingdom. Royal Commission on Environmental Pollution. 2007. *The urban environment*. London.

Vikstrom, P. 2005. 'Energy efficiency and energy demand: A historical CGE investigation of the rebound effect in the Swedish economy.' Umea, Sweden: Department of Economic History, Umea University.

Wall, G. 2004. 'Exergy.' In *Encyclopaedia of Energy*, vol. 2, ed. C.J. Cleveland. Amsterdam and New York: Elsevier Academic Publishers.

Washida, T. 2004. 'Economy-wide model of rebound effect for environmental efficiency.'In *Proceedings of International Workshop on Sustainable Consumption*. University of Leeds.

Wilson, A., and J. Boehland. 2005. 'Small is beautiful: U.S. house size, resource use, and the environment.' *Journal of Industrial Ecology* 9 (1–2): 277–8.

Winther, B.N., and A.G. Hestnes. 1999. 'Solar versus green: The analysis of a Norwegian row house.' *Solar Energy* 66 (6): 387–93.

Wirl, Franz. 1997. *The economics of conservation programs*. Dordrecht, The Netherlands: Kluwer.

Worrell, J.A., M. Ruth, H.E. Finman, and J.A. Laitner. 2003. 'Productivity benefits of industrial energy efficiency measures.' *Energy* 28: 1081–98.

Zarnikau, J. 1999. 'Will tomorrow's energy efficiency indices prove useful in economic studies?' *Energy Journal* 20 (3): 139–45.

Zarnikau, J., S. Guermouches, and P. Schmidt. 1996. 'Can different energy resources be added or compared?' *Energy* 21 (6): 483–91.

14 A Discussion of Policy Tools for Increasing End-Use Electricity Efficiency

MARK JACCARD

The chapters by Loren Lutzenhiser of the United States and Steve Sorrell of the United Kingdom contrast the lessons gleaned from more than three decades of electricity efficiency policies. While both would agree that electricity efficiency policies have not been as successful as originally promised, their chapters nonetheless present different conclusions and recommendations.

Lutzenhiser represents a school of researchers who suggest that policies to promote electricity efficiency can be greatly improved by focusing more carefully on human choice and behaviour. The design of most efficiency policies has relied all too frequently on a caricature of firms and consumers as simple cost minimizers with scarce information resources. If this were true, the provision of information and some subsidies should help to convince early-adopter firms and households to acquire efficient technologies, and soon many others would follow. However, the dissemination of most high efficiency technologies has been much slower than anticipated when these policies have been relied upon. For Lutzenhiser, an important reason is that information and subsidy programs need to be much more sophisticated. They need to be based upon careful, detailed research into the social and market context within which humans make decisions (for example, the influence of the supply chain for acquiring energy-using technologies, and the social norms within which different decision makers operate). Moreover, periods of crisis (such as the California electricity supply crisis of 2001–02) can provide additional insights into when people may be ready to make behavioural changes in order to significantly reduce their electricity use.

In contrast, Steve Sorrell represents a school of researchers who focus on the ways in which improvements in energy efficiency (including electricity efficiency) are offset, in some cases substantially, by the rebound effect. This is defined as a partially offsetting increase in energy use resulting from the lower operating costs of more efficient devices (direct rebound) and the economy-wide energy productivity gains that, in a general way, link economic growth with greater use of energy. Sorrell's conclusion is that efforts to improve energy efficiency without also increasing the price of energy are likely to be less effective than initially assumed and indeed may, in exceptional circumstances, even lead to greater economy-wide energy use in the long run. Thus, information programs, subsidies, and even efficiency regulations are compromised by the rebound effect if electricity prices are not also increased. In some cases, the price changes can be achieved within the electricity sector if the regulator or the regulated electricity providers are encouraged to implement rate designs that reflect the higher generation costs associated with new plants or the peak operation of existing plants. In some cases, the price changes will result from the introduction or intensification of economy-wide charges on environmental threats (like a rising tax on greenhouse gas emissions).

I must confess that my reading of the literature leads me to more readily accept the main thrust of the evidence and arguments presented by Sorrell. One direction of my own research is focused on all of the new devices that humans acquire as their income grows; some of these devices use energy directly, and virtually all require energy in their production, transportation, and distribution. Without an increase in its price, any effort to improve electricity efficiency by information, subsidies, and regulations alone seems like an exercise in futility.

This does not mean that we should ignore the research of scholars like Lutzenhiser. If we do finally start to act on the electricity price – so that it reflects the full financial cost of incremental electricity use as well as the full environmental damages and risks – then the lessons from his research could prove to be very valuable in designing information and subsidy programs that are a complement to, rather than a substitute for, electricity pricing.

PART FIVE

Inter-jurisdictional Cooperation in Achieving Energy Policy Goals

15 Introduction

STEVEN BERNSTEIN

Electricity markets increasingly operate across jurisdictional boundaries, both on the demand and the supply side. Considered analysis of how best to regulate and manage those markets frequently lags behind the actual flow of electricity. The reality of these flows adds an additional layer of complexity to the fundamental challenge that this volume aims to address: how to meet energy demand, in uncertain times, while significantly decreasing carbon emissions.

In this session of the workshop 'Current Affairs: Perspectives on Electricity Policy for Ontario,' contributors presented European and American experiences, both positive and negative, that might provide useful lessons for Ontario as it grapples with the challenge of designing policy that integrates energy and carbon markets not only within but also across jurisdictional boundaries. The experience of integrating energy markets, even without taking into account climate change, has proven challenging. The contributors identified the myriad potential pitfalls, inefficiencies, unintended consequences, and regulatory missteps, in equal measure to the modest successes. Moreover, lessons from the European experience of integrated climate and energy markets are only now emerging. While caution might therefore be warranted in drawing firm conclusions on the institutional and regulatory way forward, the need for attention to multi-jurisdictional regulatory, market, and institutional development could not be clearer.

Three main themes animate contributors' analysis of the institutional and regulatory challenge when policies and markets increasingly must operate inter-jurisdictionally: (1) globalization and internationalization/regionalization; (2) liberalization; and (3) surprises. First, contributors identify processes of globalization, that is, the integration of energy markets and other forms of what Jan Aart Scholte has called 'supraterritoriality' (Scholte 2005) where borders seem to matter less for flows and transactions. Coupled to this is internationalization or regionalization of policy, the idea that policy may no longer be made within traditional jurisdictions but is influenced, coordinated, or possibly even jointly constructed across jurisdictional boundaries or in transnational spaces with mixes of public and private regulation. (On the distinction between globalization and internationalization of policy, see Bernstein and Cashore [2000].) Second, the question of the balance between markets and regulation is addressed through an examination of experiences with liberalization. Third, the theme of surprises emerges, which resonates with the history of electricity policy in Ontario. I discuss each in turn below with reference to the next chapter, by Ulrik Stridbaek, and the oral commentary by the session discussant Tom Adams. In addition, I draw upon the presentation of Branko Terzic, who also provided detailed insights on these issues at the workshop from which this volume developed.

Globalization and Internationalization/Regionalization

As Ulrik Stridbaek's chapter points out, 'environmental concerns are global ... natural gas and coal markets are globalizing' and, in part as a result, 'electricity sectors are being tied together across jurisdictions, countries, and regions to an extent not seen before.' Whereas Ontario has some experience with this kind of globalization (or perhaps more accurately, regionalization), so far it has had little experience with responding to inter-jurisdictional environmental policy that interacts with those markets. European states, on the other hand, have experience both with deeply integrated electricity markets and institutional linkages as well as with the overlay of the European Union's Greenhouse Gas Emission Trading System (EU ETS) and robust national and regional climate change policies. If recent policy pronouncements are taken at face value, the situation in Ontario may soon move in the European direction. For example, the Ontario government announced in 2007 that it would join the Regional Greenhouse Gas Initiative, a partnership of eastern U.S. states to develop a mandatory cap-and-trade program to reduce greenhouse gases, and a partnership with Quebec to develop a cap-and-trade system that would link to other systems. In addition, it joined the Western Climate Initiative, a partnership of seven U.S. states, British Columbia, Manitoba, Ontario, and Quebec, to develop regional initiatives to combat climate change. Meanwhile, the Stephen Harper government has begun work to institute a system at the federal level and has also pushed for a North America–wide emissions trading system in conjunction with NAFTA.

These developments in North America and Europe suggest a clear trend toward the increasing regionalization and internationalization of environmental concern, at least in the OECD. The recent election of a government in the United States that is more sympathetic to coordinated action on climate change further suggests that existing initiatives at the sub-national level (states and provinces) may soon interact with policy initiatives at the federal level in the United States.

Looking to jurisdictions leading in this regard, Stridbaek argues that the lessons learned from Europe's experience are generally positive. For example, he notes that despite its flaws and growing pains, the EU ETS has emerged as the primary instrument to meet Europe's greenhouse gas reduction targets. Ontario's planned phase-out of coal, combined with its announced participation in regional carbon markets, is similarly likely to be the main means to achieve the Province's goals. The EU experience shows that markets can be a powerful force for international cooperation. In addition, it shows that markets can react very quickly to supply and demand, showing their potential to provide strong incentives for fuel switching or other forms of abatement.

Stridbaek draws from the Nordic experience a number of lessons for designing energy policy with these multiple goals in an internationalized setting. His basic thesis is that policy and market harmonization across jurisdictional boundaries is a necessary condition for relieving upward pressure on energy costs. Even considering the current economic uncertainty, looming investment cycles and environmental constraints will likely increase consumer demand for least-cost approaches and the need for a political commitment to see through needed policy changes to recognize these multiple global pressures.

Branko Terzic's presentation at the workshop reinforced the importance of policy coordination in the North American context, especially given the sheer size of the electricity grid. As he put it, 'the synchronized North American grids are the largest computer, the largest mechanical device [ever] built. So having some shared goals and objectives and co-coordinated policy ought to reduce the risk of unintended consequences and enhance the ability of having policy success.' For him, the core concern is the *seams* between jurisdictions. Seams exist along three dimensions: the flow of electricity, the flow of money (to stimulate investment where it is needed, reward more productive entities, and stimulate efficiency), and the flow of regulations (that is, between regimes with different market rules and designs, operating and scheduling protocols, or other control area practices). While the first two may appear straightforward, addressing the third is absolutely necessary for the smooth flow of electricity and money. Examples of policies that create friction at the seams include different pricing models, inconsistent transaction submittal times, variations in transmission tariff services, different operating criteria, and different contractual criteria. Addressing the seams is a question of interregional and international coordination and cooperation, not something that can be managed within a single jurisdiction or organically through a naturally arising market. The experience of California is a cautionary tale in this regard. That state's unique energy policies led to a problem at the seams because it imported huge amounts of power from Nevada, which did not play by California's rules. Terzic put it simply: 'If you design a market badly, it'll work badly.' Getting the rules right is absolutely essential for efficient markets.

Such cross-border harmonization and management, let alone institution building, is weak or absent in the case of Ontario. As an illustration, Terzic pointed out that the independent system operator in New England regularly posts seams issues, including those with Ontario, on its website (ISO New England). More generally, as he put it, '[a] smart grid, distribution, [and] supply choices are all considerations that have to be made in the context of the regional or international market that you're dealing with, and so the plea here is that all of these policy goals impinge upon how you in Ontario and Canada also deal with the same policy goals – hopefully comparable policy goals – across a border in which you want to trade and need to trade energy and electricity. And the synchronization, the over-the-border-looking, the co-ordination, needs to be there.'

Stridbaek maintains that Nordic countries offer a good example of mostly getting it right, although his findings also illustrate the potential and pitfalls of cross-border policy harmonization, especially in his discussion of the challenge of operating across borders in conditions of market turmoil. In that regard, he focuses on the Nordic countries' experience in 2002–3. He argues that the Nordic market was buffered against global energy market turmoil – specifically the collapse of Enron – because of its well-developed management of counterparty risk. Contemporary market turmoil, both directly in terms of commodity prices and indirectly in financial markets, only reinforces the importance of the need for careful attention to the way in which such markets are regulated. This is especially true if there is a high degree to which they depend for their smooth operation upon secondary and tertiary trading markets with high liquidity.

In sum, if we accept that globalization is a given, the demand it creates for greater harmonization and coordination, and even cross-border governance, is relatively uncontroversial. Contributors nonetheless recognize that such agreement in principle still leaves the difficulty of properly designing the means of coordination, the institutions, and the ways to harmonize policies that must respond to myriad policy interests that may be configured quite differently across multiple interacting jurisdictions.

Liberalization

More contentious than the need for cross-border coordination or harmonization, however, is the degree, speed, or means of liberalization. A reading of the EU and U.S. experiences suggests that there is no easy or universal formula for getting the mix of liberalization and regulation right.

One point of agreement is that, historically, privatization has produced significant productivity and efficiency gains. However, drawing from the U.S. and British experiences, Terzic argued that following privatization, the further unbundling of generation, transmission, and distribution only produces incremental gains. Moreover, such gains from unbundling depend on appropriate regulation. Stridbaek is more unequivocally enthusiastic about the gains from successive waves of liberalization in the Nordic region. He also makes the case that liberalized and competitive markets are good for the environment because they can incorporate the price of carbon, although it is still early days for the ETS. In this regard, the Nordic market can be viewed as an experiment in progress. In terms of Ontario, Terzic's main advice was that if one wants to tip the balance toward more competition, larger markets are better. Larger markets reduce the risk of market manipulation and market control.

Surprises

Tom Adams' commentary at the workshop added a useful cautionary note, reinforced by the other contributions. He argued that any policy is vulnerable to surprises, and Ontario has had its share. He cautioned against a herd mentality in following policy trends and urged policy-makers to try as much as possible to look to the challenges ahead rather than draw uncritically on current trends elsewhere. Stridbeak's observations about the Enron scandal and its effects in North America and the drought in the Nordic region can be combined with Adams' Ontario examples to suggest that surprises and non-linearities in policy outcomes are to be expected. Adams' comments were delivered in June 2008, before the full extent of the financial crises was upon us, making them seem even more prescient now, when the fallout for electricity policy is still far from certain.

REFERENCES

ISO New England. *Seams issues*. Available at http://www.iso-ne.com/regulatory/seams/index.html (accessed 9 December 2008).

Bernstein, Steven, and Benjamin Cashore. 2000. 'Globalization, four paths of internationalization and domestic policy change: The case of eco-forestry in British Columbia, Canada.' *Canadian Journal of Political Science* 33 (1): 67–99.

Scholte, Jan Aart. 2005. *Globalization: A critical introduction*. 2nd ed. London: Macmillan Palgrave.

16 The Power of Trade

ULRIK STRIDBAEK

Aging power plants, increasing demand, and the need to replace existing capacity with cleaner sources has lead the countries of the Organisation for Economic Co-operation and Development (OECD) into a new investment cycle. New capacity corresponding to at least 25% of existing capacity will need to be built in OECD countries by 2015. According to International Energy Agency (IEA) reference projections, of the 2,360 Gw of installed capacity in 2004 in OECD, 207 Gw will need to be replaced by 2015 and 466 Gw will need to be added to meet increasing demand (International Energy Agency 2008).

Environmental constraints are constantly tightening, and ambitious environmental targets have been set by governments. Innovative policies are needed to enable these challenges to be met at reasonable costs. Considering that many of the environmental concerns are global, and considering that natural gas and coal markets are globalizing, electricity sectors are being tied together across jurisdictions, countries, and regions to an extent not seen before. Currently, increasing gas and coal prices are having effects on the costs of power generation. Spot fuel prices represent an opportunity cost even for the power generators that have long-term gas and coal contracts at lower prices. In competitive power markets the increasing fuel costs will tend to be passed on to wholesale power markets and eventually to final customers.

Enabling the electricity sector to adapt dynamically to these trends and in general to benefit from optimized resource allocation across

---

This chapter is based on various International Energy Agency publications, including IEA 2005, 2007a, 2007b, and 2008.

larger areas is an opportunity and most likely a necessity. Tighter fuel markets, tighter environmental constraints, and a looming investment cycle put strong upward pressures on electricity costs. It is essential that these pressures are eased as much as possible through least-cost approaches and optimization to win public acceptance and the political commitment to create a framework that allows for the critically necessary transformation of the electricity sector. To that end, innovative policies and harmonization across jurisdictional borders are required.

Competition as an Instrument for Transformation

The introduction of competition and regulatory reform provides strong instruments for the necessary transformation at least cost. Liberalization is on the agenda in all OECD countries and in many non-OECD countries. In some markets where reform has failed, the merits of reform are seriously questioned, for example in some states of the United States where reform has now rewound. Several other markets have liberalized with considerable success, showing that it can be done but also that it is a long process that requires ongoing government commitment.

Over the last ten years, markets have been liberalized and have become robustly competitive in several places, for example, Australia, the Nordic countries, the United Kingdom, New Zealand, and in Texas and several northeastern U.S. states.

Maintaining robust investment incentives in liberalized markets is one of the most contested issues of reform. All markets have some measures in place that put a price on electricity capacity, thereby creating price signals that are additional to the pure energy price. In some markets, like in Australia, the capacity element is at a minimum, with a focus on reserving capacity to deliver ancillary services. This is a one-price-only market where the megawatt-hour price is intended both to cover marginal costs and to give remuneration to invested capital. Other markets, like in several U.S. states, have chosen to operate two parallel markets. The one-price-only markets naturally have considerably higher price volatility, and they have, so far, proven to give robust investment signals. For example, in Australia electricity prices have triggered investment in peak units, and in Finland the current investment in a new nuclear unit is based on expectations of future average Nordic electricity prices.

The internalization of external environmental costs – for example, targeted by a cap-and-trade system for carbon dioxide emission permits – adds considerable uncertainty for investors. The effects of greenhouse gases on climate change, and not least the political responses to these challenges, add inherent risks to all investments. Uncertainty about the future framework forces investors to hold back investments and to divert investments to less risky opportunities such as combined cycle gas turbines. Introducing capacity targets to secure adequate investments is not the appropriate response to the challenges posed by these politically driven risks. Capacity targets would accelerate investments, risking locking in investments that may prove to be inappropriate. Policymakers should instead concentrate efforts on reducing the risks inherent in the climate policy, and speed up making the necessary decisions on the critical framework conditions.

Liberalization as a Corner Stone in European Union Energy Policy

Continued liberalization is a central piece of European Union energy policy. It is regarded as an essential prerequisite to meeting EU energy policy commitments and goals. A new, comprehensive energy policy package was agreed by European ministers in the European Council in March 2007. Greenhouse gas emissions are to be reduced by 20% by 2020; 20% of total final energy consumption is to come from renewable energy by 2020; and energy consumption is to be reduced by 20%, also by 2020. The greenhouse gas emissions reduction target is increased to 30% if a global abatement agreement can be reached. In addition, it was agreed to step up efforts to create an internal energy market. Increasing concerns about security of energy supply and European competitiveness due to increasing energy costs were some of the key drivers, in addition to environmental concerns. The European Commission has, since then, made concrete proposals on how to implement the agreed policy package. On liberalization, the European Commission has proposed a third liberalization package; the first was agreed for electricity in 1996 and for gas in 1998; the second liberalization package was agreed in 2003.

The European Union embarked on a lengthy liberalization process in the mid- 1990s. The second market directive included unbundling provisions, a provision to give all EU electricity and gas customers freedom to choose their supplier from 1 July 2007, and relatively precise provisions of the principles for cross-border trade. Annual benchmarking reports and a 2007 competition sector inquiry have revealed great advancements in the introduction of competition in some markets, but also considerable delays and inadequacies in many others. This triggered the third liberalization package with stronger measures proposed, notably on unbundling of transmission system operators, strengthening of regulatory functions, and greater coordination between national regulators and system operators. Enhancing transparency and protecting consumers are also key issues. The details of this package are being discussed now, with a focus on how far to go in unbundling provisions and what power to give to a new European regulatory agency.

The European Commission has proposed to require full ownership separation of transmission networks and transmission system operation, accepting an ISO model as a second best. The ISO model would allow transmission grids to be owned by vertically integrated companies but leave full control of networks and network planning with a fully independent system operator. European governments are radically divided on the need for such strong measures, with a minority, including Germany and France, strongly opposing.

The European Union Greenhouse Gas Emission Trading System is the corner stone of EU policy to abate greenhouse gas emissions. The first phase was launched on time, as planned, in 2005. A second phase started in 2008. European Union governments have agreed to extend the system beyond the second phase, which ends in 2012. The EU ETS is the main tool to be used to meet the 20% greenhouse gas reduction target by 2020. The decision to extend the EU ETS is founded on the conclusion that even if the lessons so far have highlighted weaknesses and room for improvement, the system has been shown to be a very powerful instrument for international cooperation on solving international problems.

The price of CO_2 emission allowances plummeted from about US$35/ton CO_2 in the first months of 2006 to roughly US$0 a year later. Most of this price drop happened during the last three months of 2006 in connection with the collection of official emission statistics that showed that emissions were considerably lower than expected. Transparency and data management were highlighted as being of critical importance during these events. It also clearly illustrated that markets can react sharply and forcefully based on assessments of supply and demand, showing the potential powers of the instrument if used carefully.

According to analysis by Fortis Bank, the EU ETS created incentives that led to fuel-switching from coal to gas, which resulted in abatement of 88 metric tons CO_2 in 2005, and 59 Mt in 2006. The 88 Mt of CO_2 corresponded to 6.1% of all CO_2 emissions in the EU electricity sector in 2005 and to 4.1% of CO_2 emissions in all ETS sectors.

Some of the most important tools enabling trade and cooperation on environmental constraints are now in place or are developing in the European Union. A key challenge for European Union at present is to let the internal electricity market improve further. Experiences in the Nordic countries have been a point of reference and a positive example for the development of an internal European market.

Competitive Markets Tie Nordic Countries Together

The Nordic market evolved with stepwise market opening in Norway, Sweden, Finland, and Denmark. All electricity customers in all Nordic countries have been free to choose a supplier since 2003. There are three main pillars in the Nordic market: transmission system operators (TSOs) are fully separated from generation and retail supply; regulators are strong and independent; both TSOs and regulators cooperate across all Nordic countries and work continuously on harmonizing rules and regulations.

The development of the Nordic power exchange Nord Pool is one of the key points in the harmonization efforts. Nord Pool operates a day-ahead spot market. It is a voluntary market, and liquidity has improved continuously, recently reaching a market share of more than 70% of total Nordic consumption. Nord Pool also operates a market for financial derivatives and a clearing service for financial contracts traded over the counter. Liquidity in financial contracts improved slowly but steadily from its launch in the mid-1990s, with a considerable setback in 2003 in connection with serious market turmoil. Liquidity started to increase again and is now above previous levels, corresponding to about eight times the total Nordic consumption. The array of standardized contracts traded at Nord Pool has increased steadily, and today contracts can be traded five years ahead of time.

In addition to the day-ahead and financial trade at Nord Pool, the Nordic system operators operate a common real-time market, allowing flows across country borders to be optimized also in real-time balancing. Nord Pool operates an intra-day market, allowing for trades between the closure of the day-ahead price settlement and the closure of the gate for real-time system operation. Liquidity in the intra-day market is very low.

The Nordic market turmoil in 2002–3 was unrelated to the debacle that had taken place in many other markets in the previous years as a result of the collapse of several U.S. utilities. The Nordic market had been more or less unaffected by the collapse of one of the larger energy market players, Enron, because of relatively developed management of counterparty risk. The collapse of Enron highlighted the usefulness of clearing services for over-the-counter trades. The Nordic market turmoil was caused by a serious drought in the Nordic region and low levels in hydro reservoirs. The electricity system has a very high dependency on hydro power. Inflow to Norwegian hydro reservoirs was far below normal, depriving the Norwegian system of 18 terawatt-hours during the second half of 2002, which corresponded to 15% of Norwegian consumption in 2002. Figure 16.1 shows the level of Norwegian and Swedish hydro reservoirs, measured in terms of deviation from normal.

The Nordic system had lost 35 terawatt-hours by the end of 2002, corresponding to 9% of total Nordic electricity consumption in that year. The market response was swift. Prices quadrupled, triggering marked responses from all sources of flexibility. Thermal power production and imports were maximized. Customers also responded by decreasing demand by as much as 5%.

Market prices directed power flows across country borders. Prices directed relatively large exports from Norway to Sweden in the second half of 2002. The drought had already started to make its mark in Norway, but the drought in Sweden was even more severe. There were considerable tensions and public debate, particularly in Norway, but governments resisted the temptation to intervene. After the end of the drought several reports and studies, including a report commissioned by the Norwegian parliament, confirmed that the market had optimized the use of very scarce resources through the event. Competition and trade had critically helped to solve the situation in the best possible way, rather than being the source of the problem as claimed publicly by some decision makers.

The Nordic market is also a good example of how liberalized and competitive markets can be a useful tool to meet environmental goals. Denmark has been one of the leading countries in the development of wind power and is still the country with the highest share of wind power in the world, reaching 20% of total electricity demand. On some occasions the western part of Denmark, with the highest concentration, experiences wind power generation exceeding total demand. The Nordic power market has proved to be an ideal tool to overcome the challenging task of integrating wind power and managing its variability. Figure 16.2 shows wind power generation, consumption, and flows across the interconnectors with Norway and Sweden.

Figure 16.1. Hydro Reservoir Levels in Norway and Sweden (Percentage Deviation from Normal), Weekly Average Nord Pool Prices, and Trade between Norway and Sweden

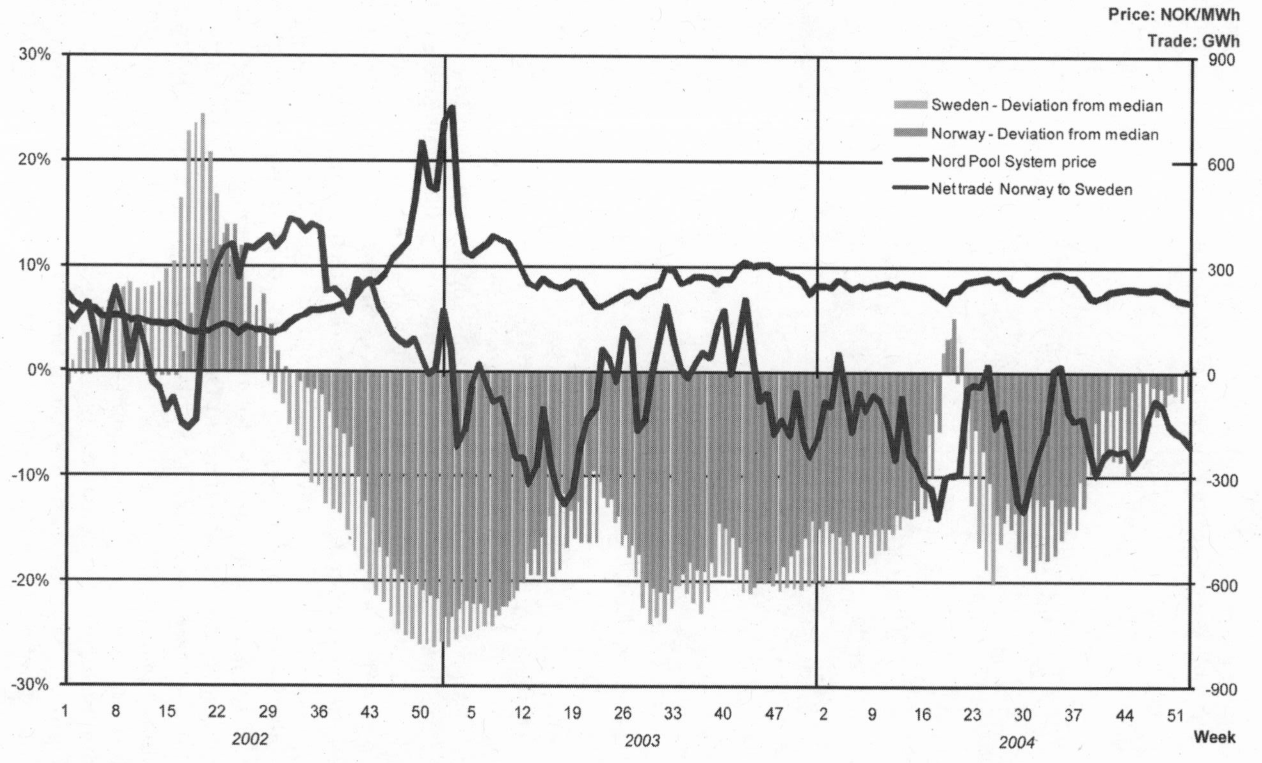

Source: Nord Pool, adapted from ECON Energy 2003.

Figure 16.2. Western Denmark's Demand, Wind Power Generation, and Flows across the Interconnectors with Norway and Sweden, December 2007

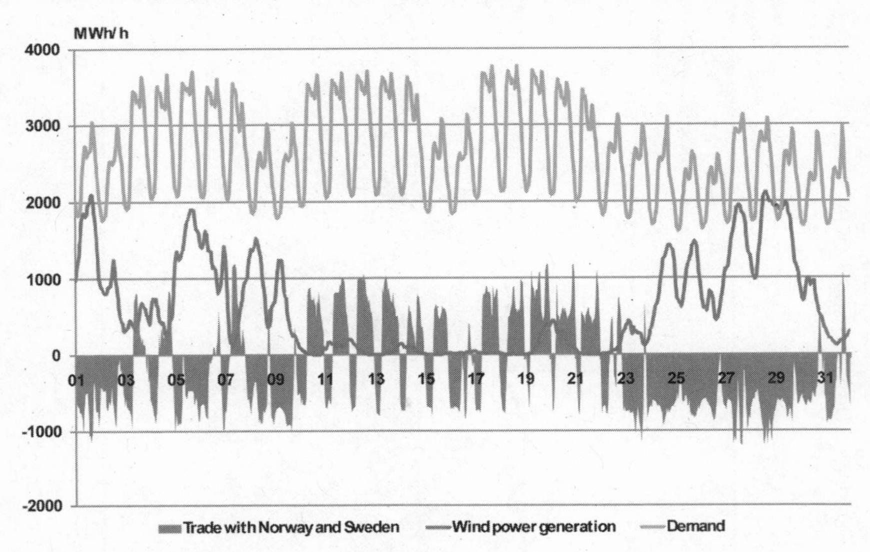

Source: Energinet.dk

The data in figure 16.2 shows that power tends to flow to Norway and Sweden when wind generation is high. On 29 December 2007 wind generation in western Denmark was higher than the total demand. High shares of combined heat and power for district heating add another must-run source to the system that can overflow the western Danish electricity system with power during cold and windy periods. Figure 16.3 illustrates the price reactions for the same period.

Figure 16.3 shows day-ahead Nord Pool prices for the entire Nordic system – the so-called system price – and for western Denmark. The Nordic market uses a model for zonal pricing, usually with seven Nordic zones, two of which are in Denmark. The system price represents the unconstrained price, the price resulting if there were no transmission constraints in the Nordic system. This price is used as the main reference price for financial contracts. The price for western Denmark is the price derived when transmission constraints are taken into account. Figure 16.3 illustrates that western Denmark can be an area with prices considerably higher than the Nordic price. This is related to the role

Figure 16.3. Wind Power Generation and Day-Ahead Nord Pool Prices for the Nordic System and Western Denmark, December 2007

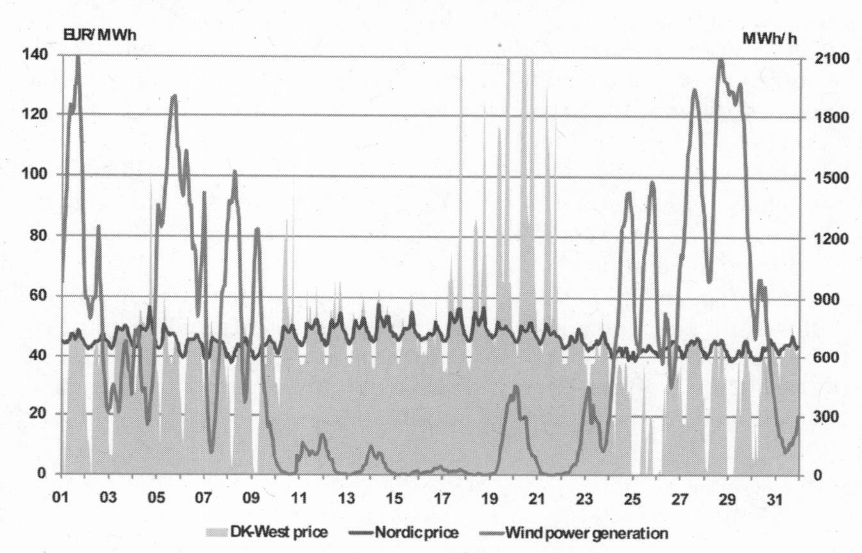

Source: Nord Pool and Energinet.dk

that Denmark play as a gateway between the Nordic hydro-based system and the continental European thermal-based system. The picture often changes when wind power generation is strong. During several hours in the last week of December 2007, strong wind generation and relatively low demand drove the price to zero – which has happened frequently during the past five to seven years.

Even if the Nordic market is mainly dependent on hydro and nuclear power, fossil-fuelled thermal plants are most often on the margin, setting the price. The Nordic market is thereby also a good testing ground for the impact of the EU emission trading system. The price of CO_2 emissions did seem to be passed through to the power price as intended, lifting it to a higher level and thereby creating additional incentives to switch away from CO_2 intensive fuels both in operation and in investment. Construction of a new nuclear reactor was started in Finland in 2004, the first new reactor in an OECD country outside of Asia for almost two decades. Power prices with internalized CO_2 costs are one of the contributing factors to the economic feasibility of that project.

Key Messages

Liberalization is a long process that has well served those markets which took early and comprehensive reform actions. Comprehensively liberalized markets have delivered efficiency gains and transparency. They have so far proven to give robust investment incentives, offering market participants real measures to manage risks. Finally, the dynamic responsiveness created in the competitive framework of liberalized markets will be essential in meeting the requirements of transformed and cleaner electricity systems, with more volatile and more widely distributed generation.

The European Union strengthens its market reform efforts in parallel with setting up more ambitious environmental goals, seeing competitive markets as an essential instrument in undertaking the necessary transformation of the energy sector without jeopardizing European competitiveness and security of supply.

REFERENCES

International Energy Agency (IEA). 2005. *Lessons from liberalised electricity markets*. Paris: OECD/IEA. http://www.iea.org/Textbase/publications/index.asp.

– 2007a. *Tackling investment challenges in power generation*. Paris: OECD/IEA.

– 2007b. *World energy outlook 2007*. Paris: OECD/IEA.

– 2008. *Energy policies of the European Union*. Paris: OECD/IEA.

PART SIX

Policy Challenges and Opportunities

17 Institutions Matter

MICHAEL J. TREBILCOCK

In analyzing the public policymaking process in almost any context, I have always found it helpful to contemplate the following sequence of steps: (1) institutions, (2) instruments, and (3) objectives. Presumably, in a rational public policymaking process, a given set of institutions would determine collectively the desired policy objectives and would then choose an appropriate set of policy instruments to vindicate those objectives.

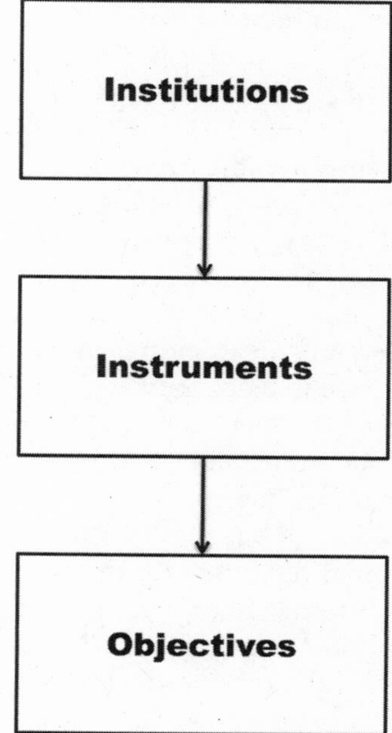

At the Current Affairs conference, both in the address after dinner and in the presentations during the following day, most of the discussion focused on the appropriate choice of policy objectives and the appropriate choice of policy instruments to vindicate those objectives. Very little discussion focused on the existing or prospective institutions that might determine the choice of objectives and the choice of instruments. In my approach to these issues as a lawyer and political economist, the key questions in this public policymaking context (and most other such contexts) are: Who gets to decide and by what process of decision making? Or who is herding the cats? Or who has the hand on the switch (or the tiller)?

In the field of economic development, a paradigm shift has occurred over the past decade or so. Of all the potential determinants of a country's economic development trajectory, a consensus is emerging that the quality of a country's political, bureaucratic, regulatory, and legal institutions is a major determinant of that trajectory – if not the most important determinant. This view has now been substantiated by a rapidly growing body of empirical literature that emphasizes that the quality of a country's institutions that formulate, implement, administer, and enforce public policies is of crucial importance. Yet curiously, in the conference, the predominant focus on policy objectives and policy instruments, and critiques of current policy objectives for incompleteness or incoherence, or of current choices of policy instruments as inappropriate or ineffective, largely ignored the importance of analyzing the institutions that have generated or are likely to generate the choices of policy objectives and choices of instruments.

In the energy and electricity fields there is a veritable cornucopia of institutions. At the supra-national level, multilateral institutions are emerging to monitor compliance with international obligations under treaties such as Kyoto I and II. Regional institutions are emerging to manage cross-border trade in electricity and other forms of energy.

At the national or federal level in Canada there are separate ministries of energy and the environment. There is also the National Energy Board, which regulates interprovincial and international energy flows, although to this point it has played a largely inconsequential role in promoting and regulating interprovincial and cross-border flows of electricity. There is Atomic Energy Corporation Limited, a Crown corporation, which manufactures the CANDU reactor. There is the Canadian Nuclear Security Commission, which regulates the nuclear industry in Canada.

At the provincial level there are separate ministries of energy and the environment. There is also the Ontario Energy Board, which regulates an increasing range of issues in the electricity sector. There is the Independent Electricity System Operator (formerly the Independent Market Operator), which operates the electricity spot market and performs the dispatch function. There is the Ontario Power Authority, which contracts for the production and supply of electricity from independent generators. There is Hydro One, an entity owned by the Province of Ontario, which operates the transmission grid and presumably performs various public policy functions that an investor-owned, profit-maximizing entity may not (otherwise it would be one). Similarly, Ontario Power Generation, which operates 70% of the generation capacity in the province, is owned by the Province and also, by assumption, performs public policy functions that a privately owned, profit-maximizing entity may not.

At the municipal level there are some eighty or so municipally owned local distribution companies – in effect, enterprises owned by the Province – that, again, by assumption, perform various public policy functions that privately owned, profit-maximizing entities (which are subject to regulatory constraints) may not (otherwise they would be such entities).

In my view, critiques of choices of current policy objectives and choices of current policy instruments are largely beside the point unless we are prepared to analyze the institutions that have generated these choices of objectives and choices of instruments. Such critiques are largely an exercise in whistling in the wind.

To return to the questions posed at the outset of this chapter, is it clear who amongst this melange or plethora of institutions gets to decide critical choices of policy objectives and choices of instruments? One might hypothesize that amongst these institutions the primary candidate for lead herdsman of the cats is the provincial ministry of energy. However, in this respect I have misgivings. Relatively rapid turnover of ministers and deputy ministers. and the highly attenuated internal policy analytical capacity which has been reflected much more generally in recent decades across both the federal and the provincial government, ill equips the ministry to play this role. Such policy analytical capacity would not be merely inwardly directed to providing policy advice to the minister but would engage the external research community and stakeholders and take citizens into the confidence of government in exploring policy options, best practices, and painful trade-offs in formulating choices of policy objectives and instruments. But if the ministry is not equipped to play this role, who is (particularly given the paucity of independent think tanks in Canada relative to other countries)? To leave the institutions step in the policy development sequence largely unanalyzed (no matter how messy and complex this exercise) is largely to treat it as a black box, if not a black hole, and explains largely why academics of the pre-eminence of many participating in the conference are left with a sense of frustration and puzzlement as to why their often well-taken critiques of choices of policy objectives and choices of instruments have little traction in the real world.

18 The Politics of Electricity in Ontario

SEAN CONWAY

It is a pleasure to have this opportunity to provide some concluding comments about the timely and important subject of electricity policy for Ontario. Let me say at the outset that my perspective on this topic is that of someone who has spent nearly thirty years of his life in the Legislative Assembly of Ontario. Both as a member of the Opposition and as a Cabinet minister, I devoted a good deal of that time to wrestling with what often seemed to be the endlessly complicated hydro issues. What follows are the observations of the battle-scarred!

I hope that it is to state the obvious that the supply of affordable, reliable electricity is today what it always has been – an extremely important value for Ontarians. Whether at home or in the workplace, Ontarians understand that few things are as vital to their economic and social well-being as electricity. They understand that theirs is a large, industrialized province that is very cold for a significant portion of the year. As the ice storm of 1998 reminded many of them, their personal safety is at risk if, for any reason, the supply of electricity is interrupted, especially in the winter months.

I say this because sometimes the debate about electricity misses this very essential point. This commodity is not like others, because electricity cannot be stored; it must be there when the consumer flips the switch, and any major interruption will cause significant difficulty and real hardship. Policymakers and politicians would do well to understand that it is this combination of qualities that gives electricity its peculiar and powerful salience in Ontario's political culture.

Speaking of political culture, there is one other feature that is highly relevant here. We teach students in Canadian universities that regionalism is one of the distinguishing characteristics of our political culture.

I can tell you from plenty of personal experience that electricity policy is one of those issues that will take you directly to some of the most sensitive and dangerous elements of our regional divisions. Remember that most consumers live in cities while most electricity is generated in non-urban and northern areas.

Another observation I would offer based on my experience is that the electricity debate is often driven as much by 'theology' as it is by rational analysis. Let me be more specific here by using the example of the Ontario nuclear debate of the 1970s and 1980s. Very enthusiastic supporters of the nuclear option saw it as a 'heaven-sent' solution because it would provide relatively cheap, home-grown energy from a distinctively Canadian technology in a province that always valued energy self-sufficiency. Opposing this view were equally enthusiastic people who felt that this huge nuclear commitment was wrongheaded and bound to fail. This very passionate argument about the role of nuclear power in Ontario often generated more heat than light.

The 1990s surely taught Ontarians that there were important questions to be answered about the efficacy of nuclear power in Ontario's energy future. Governments continue to struggle with some very thorny issues in the nuclear sector such as the true cost of construction and operation of these very sophisticated plants. I will not even engage the very controversial issue of nuclear waste disposal, except to say that it does remind us that today's electricity policy must meet a new environmental imperative.

In my view, it is absolutely critical that we find a mechanism for a better kind of public debate about the critical issues facing Ontario's electricity sector. It has been my experience that we have a long way to go in terms of engaging the public in a more sustained and honest discussion about the real choices we face and the costs and consequences of these choices. Let me be a little clearer. We now know that the big nuclear commitment that Ontario made in the post–Second World War era was a lot more complicated and costly than many of us initially imagined. Over the past fifteen years we have learned some shocking things about what happened as we travelled along that nuclear highway. We are now on the verge of another major nuclear investment in Ontario. My question is, are we confident that we have truly learned the lessons from our earlier experiences with this technology?

Ontarians are a practical people who care about economic prosperity and managerial competence. The electricity question invites active scrutiny on both of these fronts. We are a province and a people with few myths, and make no mistake about it, Adam Beck and his creation, the Hydro-Electric Power Commission of Ontario, represent one of the most powerful and enduring myths in our history. The 'Hydro' has delivered enormous benefits to Ontarians. Notwithstanding that a lot has changed, the citizens of this province expect that planners and politicians are going to get this right, that they are going to decide these tough questions in a way that accords with the fundamental values mentioned above.

19 Conclusion: Challenges and Opportunities for Electricity Policy in Ontario

DOUG REEVE AND DONALD N. DEWEES

We can identify three realms in which electricity policy for Ontario presents considerable challenges and many opportunities for leadership: the environment, the economy, and politics. These three realms overlap, of course. The political problems would be easier to solve if we were not concerned about the environment and if stakeholders could agree that energy conservation opportunities were either cheap and abundant or costly and limited. Governments have not told Canadians that vigorous environmental protection will impose costs on all of us and will require some lifestyle changes. Therefore, there is a lingering illusion that someone else will pay to green the economy. We need better dialogue regarding the economic and environmental implications of phasing out coal, expanding renewable supply, or building a new generation of nuclear plants, and the role of central planning versus market forces in each of these decisions. Engaging the public on these complex issues is a daunting challenge.

The Environmental Challenge

Periodically the 'environment' is the leading cause for concern among Canadians, as it was for much of 2008. In the summer in southern Ontario, air quality, which is directly related to coal-fired electricity generation, is a recurring issue. However, environmental concerns are regularly overtaken by bread-and-butter issues, and no poll is needed to predict that the financial crisis and economic slowdown that began in the fall of 2008 will keep the economy high on the agenda for the next few years. The challenge for Canadians and for our political leaders as we work through the current economic problems is not to forget our concerns about the environment and not to falter in our development of policies to meet our long-term environmental needs. We must not fail to respond to the spectre of global climate change.

In situations where we recommend policies that would lead to investment in low-carbon generation or in energy conservation and efficiency, the economic downturn provides an opportunity to direct some of the government stimulus and infrastructure spending scheduled for 2009 and beyond towards these environmentally protective projects.

Global Climate Change

As Edward Parson points out in chapter 2, there has been much debate about how far we will need to go to avert climatic disaster and how quickly we must respond. It is now clear to most people who have studied the problem that the global output of greenhouse gases must be decreased by 50%–80% by 2050. Yet, there are some who either do not believe we are on a dangerous course or do not believe it requires action on our part. The federal governments of both Canada and the United States in 2008 did not appreciate the urgency. In the United States in early 2009, President Obama does, but it remains to be seen whether or not other levels of government do. There is an expressed political will in the European Union, but little evidence of such will in important countries like China and India. If the stage is not set globally, continentally, or even nationally, what can Ontario do? We must lead by example, and we must prepare ourselves for a future when others finally swing into action. We must choose climate action, not catastrophe.

Electricity policy is inevitably central to climate change policy and to the main driver for climate change, fossil fuel (energy) consumption. That leads us to three conclusions: (1) electricity policy must be seen in the context of climate action targets to 2050, and therefore the planning horizon is much longer than usual; (2) electricity policy is not independent of energy policy but is an integral component; and (3) energy policy must address fossil fuel consumption for electricity but also for other sectors of the economy.

Policy goal for Ontario: *Climate change guidance for all energy and energy-related policy.*

The growth in fossil fuel consumption is driven in large part by growth in per capita income and growth in population. If we as a global society are to radically decrease fossil fuel consumption, we have the options to (1) find non-carbon energy sources, (2) develop more energy-efficient technology, (3) alter our lifestyles, and (4) limit global population. There is little appetite for, and limited politically feasible means of, achieving these last two, and so we tend to focus our search for technological solutions on the first and second options.

There are a number of ways in which the development and deployment of non-carbon energy can be stimulated, but it is essential that there be a price on carbon emissions (see chapter 5). First and foremost in universality, simplicity, and directness is a carbon tax. No one can doubt the difficulties of selling a carbon tax, particularly given the economic circumstances of 2008–9, and indeed it may be politically impossible at this time. An alternative is to use cap-and-trade systems to limit greenhouse gas emissions and to put a price on those emissions. This could be effective, but the complexity and technical problems of cap-and-trade systems make them a distant second best to the carbon tax as a means of pricing, and thus discouraging, carbon emissions (see chapter 4). However, with the United States moving toward cap-and-trade policy, it may be too late for a carbon tax. Either way, effective carbon pricing that makes greenhouse gas emissions expensive must be achieved if we are to maximize the prospects of achieving the climate targets of 2050. Solutions are at hand: smart politicians can find a way; the United States may move so decisively that we have to follow; or looming catastrophe could persuade us to act.

Policy goal for Ontario: *Pricing of carbon emissions.*

Electricity generation in Ontario is aimed at a low-carbon future, with the planned shutdown of the coal plants in 2014. However, gas-fired generation is currently part of the supply strategy. To achieve a low-carbon future, we will depend for baseload on nuclear and renewable energy sources, such as water, wind, and sun (see chapter 4). Nuclear reliability in Ontario will have to be much better than it has been. Recently built gas-fired plants are, in theory, to be used only for peaking; we will need to turn this theory into practice in order to achieve a low-carbon future. Storage will have to play a key role in responding to short-term demand fluctuations and time-varying renewable generation. Links with neighbouring jurisdictions can also play an important role, as we have learned from the experience of wind-generated electricity in Denmark being backed up by hydroelectric power from Norway and Sweden (see chapter 16). In assessing these technologies, we should look not just at generation emissions but also at the life-cycle environmental impacts of each technology.

Ontario has been encouraging renewable generation through its standard offer programs for wind, solar, and small hydro generation. Europe has succeeded in stimulating new renewable generation using cap-and-trade systems to make carbon emissions expensive, along with a form of renewable portfolio standard to mandate a proportion of green energy, or a feed-in-tariff that specifies a high price to be paid for green energy generation (see chapter 4). The feed-in-tariff has been much more successful in causing investment in green generation that costs much more than the market price for electricity, and Ontario seems to be following Europe's lead with the feed-in tariffs proposed in the 2009 *Green Energy and Green Economy Act*. However some renewable power is very expensive, and there is little justification for paying more than necessary for greenhouse gas reductions. The solution is to compare policies based on the net cost per metric ton of greenhouse gas reduced. We cannot afford to waste money on very high-priced carbon reduction strategies.

A related option is the generation of electricity from municipal solid waste, technology that is used in some of the most environmentally conscious jurisdictions in the world. In Ontario we recycle a great deal of solid waste, but we inventory the considerable remainder in landfills, which are sometimes hundreds of kilometres from the source. Some of the landfill waste is made up of carbon from plant matter (paper, cardboard, et cetera), not from fossil fuel, and therefore, although it is not carbon free, it is climate-neutral – based on the short-term cycling of carbon through growing plants, as long as the forests and fields are harvested sustainably.

Policy goals for Ontario: *Low-carbon electricity and coherent encouragement of economical renewable generation.*

Electricity can play a significant role in decarbonizing the transportation sector through plug-in hybrids and electric vehicles, so long as our electricity generation mix burns little fossil fuel (see chapters 2 and 6). Although public transit development is presently stalled, at least in Toronto, by 2050 perhaps we will see substantial increases in passenger transportation that is electrically powered. We might imagine a mix of fixed rail, trolley buses, hybrid buses, and perhaps independent electric buses, with the appropriate choices for the time and place made based on economics and available technology. We might also see more dense urban and suburban developments that require less driving and are better suited to mass transit. Similarly, it presently appears as though trucking will continue to grow for transportation of goods, but as the cost of fuel and of carbon emissions increases, perhaps we may see redevelopment of rail and electrification of some high-density rail lines. As home heating by gas or fuel oil suffers from similar increases in cost, so we may see greater demand for electricity for heat pumps and geothermal heating.

Policy goals for Ontario: *Low-carbon electricity that provides energy for transportation and heating; and electrically powered public transit for major urban centres.*

Air Quality

The planned phase-out of coal by 2014 will remain uncertain until conservation and alternative generation have proven that they can fill the gap. It would therefore be prudent to be concerned about conventional pollution emissions from these coal plants for more than the five years to 2014. Conventional coal combustion releases oxides of sulphur, oxides of nitrogen, particulate matter, and various persistent toxic substances such as mercury and cadmium. All of these substances cause environmental harm and pose direct health risks to humans. Technology can reduce these emissions to low levels per megawatt-hour of electricity in conventional coal-fired power plants with moderate increases in capital and operating costs. Alternatively, integrated gasification combined cycle plants or 'clean coal' technology can reduce the air emissions to very low levels. This technology is costly, but it may cost less than the competing clean energy sources, including some forms of renewable energy. Regulations that require very low emission rates from fossil-fuelled power plants will let generators choose whether to close their coal plants or to fit or rebuild them with appropriate technology that meets our environmental goals.

Policy goal for Ontario: *Virtual elimination of pollutant emissions from coal-based electricity generating stations in Ontario.*

Nuclear Power

The government of Ontario has made it very clear that nuclear power is to provide baseload electricity well into the future, and some people believe that it is a necessary feature of future supply. Nuclear power is not a significant contributor to greenhouse gas emissions except when one looks at the life cycle of the generating station and its fuel; if electricity from nuclear power is used to supplant fossil fuels from transportation and heating, as discussed above, nuclear power will be an even greater contributor to a low-carbon (but not zero-carbon) future. Nuclear power does not contribute to particulate emissions and poor air quality directly, as coal does, but there are environmental concerns about construction of the plant and the mining and production of the fuel. Moreover, there is ongoing concern about the environmental risks of nuclear power, the operating safety of nuclear reactors, the risk of theft of nuclear materials for non-peaceful purposes, and particularly the storage of waste (see chapter 2). Resolution of the waste storage issues should be an important part of the government's energy policy.

There is also concern that past nuclear investments, for new plants and the rebuilding of shut-down units, have experienced long delays and high costs. The reliability of Ontario's nuclear units has been disappointing. Nuclear economic performance elsewhere has been distinctly mixed. Before we invest in new nuclear projects, we need some assurance that costs, delays, and reliability will be acceptable, yet such assurance will require actual construction and operating experience somewhere. There is no operating experience anywhere with the advanced reactors that are being considered for construction in Ontario. If Ontario orders new reactors in the near future, it will be doing so at the same time as other utilities are placing orders, thereby raising risks of cost escalation. We need frank and realistic dialogue about the way in which these risks will be shared between electricity consumers and the investors in new plants and about the extent to which taxpayers will subsidize nuclear electricity through government backstopping of such projects.

Policy goals for Ontario: *Safe, secure, long-term storage of waste from nuclear power generation; and careful evaluation and discussion of nuclear technology as to cost, environmental performance, economic performance, and the risks faced by the public, electricity consumers, and taxpayers.*

The Economic Challenge

Economic Policy for the Province

The Province of Ontario aims to provide an economic environment that offers a satisfactory standard of living for all Ontarians. The connection between this goal and electricity policy, however, is not strong. A sound economy requires a reliable supply of electricity that is adequate to meet our needs at a price that covers all costs, including environmental harm, and that is not damaging to the economy (see chapter 8). In the past, cheap electricity has been used to attract industry to Ontario, which was a wise policy when we relied predominantly on a low-cost source of supply, the Niagara River. Now that all our newly built and future generation is of the same type and can be built by our competitors in the Great Lakes region, we have no cost advantage. It would be a mistake to ask all taxpayers in Ontario to subsidize cheap electricity in order to attract or retain industries that are highly electricity intensive. If our goal is jobs and we want the government to intervene, we will get more bang for the buck by subsidizing the jobs directly. The goal for electricity pricing should not be to subsidize but simply to ensure that electricity consumers pay all the costs of the electricity system including the cost of environmental harm. This provides a level playing field between electricity and other forms of energy and does not encourage energy-intensive industry to locate in Ontario, thus forcing us to build more high-cost generation.

Policy goals for Ontario: *Electricity infrastructure that provides a reliable and sufficient supply of electricity to meet demand; and payment by electricity consumers of all the costs of electricity including environmental harm.*

Electricity Supply, Delivery, and Demand

Planning electricity supply is a complex business: investments are large and long term; return on investment is complicated; a dedicated, instantaneous delivery system is required; many levels and departments of government are involved; and many approvals are required. Planners also have the challenge of estimating demand, which is another complex business: the population is growing; companies come and go; conservation programs aim to increase efficiency and thereby decrease demand, but often there is a rebound effect (see chapter 13); and the growth (or shrinkage, as recent events suggest) of the economy is difficult to predict. The planning of time scales must take into consideration not just years but decades, and now, with the need for response to climate change, we have to plan for half a century. One thing is certain: the target – reliable, robust, cost-competitive supply that is sufficient for demand – is a moving target.

Ontario's neighbours to the south, east, and west also generate and consume electricity, and linking our grid to theirs provides additional dimensions for supply and for demand. At times, Ontario imports from Quebec, a province with abundant hydroelectric capacity, and exports to the United States, a large economy with growing demand. Is electricity self-sufficiency a desirable policy? Is electricity supply a growth industry for Ontario?

Given the stability of our neighbours it seems unnecessary to be entirely self-sufficient, and therefore, if our neighbours have cost advantages and excess capacity, we should invest in their facilities and/or buy from them. If we have cost advantages and excess capacity, we should sell to them. Ontario does not hold a superior position relative to our neighbours with respect to new electricity supply for the future: we have no advantageous coal resources or carbon sequestration geology; we have limited and faraway hydroelectric expansion prospects; the sun does not shine more brightly in Ontario, nor does the wind blow more fiercely; and our record on nuclear power does not suggest that we have a competitive advantage with nuclear, in spite of massive Canadian investment in nuclear technology. Nor do we appear to have a technological advantage in any other major energy technology. All that being said, we may not be a big electricity exporter in the future, but we do not have the cost disadvantages that suggest we should plan to become a major net importer. Until there are changes in these relative costs, we can best secure reliable, environmentally responsible, and reasonably priced future supply with predominantly domestic generation.

Long-range transmission is also an important issue for the future. Demand growth is in southern Ontario, and new renewable generation, by wind and water, is in northern Ontario or perhaps beyond our borders. Planning, financing, and constructing transmission lines is a straightforward exercise except for the cost of building and the complications of locating the lines. As will be discussed in greater detail below, there are substantial political constraints to be overcome.

Apart from the long-term Canadian investment in nuclear technologies, there is not much to be said about Ontario as a world leader in electricity or energy technology. As a global society we will need to achieve exceptional reductions in carbon emissions and, where there is no carbon capture and storage, in fossil fuel consumption. An important alternative source of energy will be low-carbon electricity supply and the systems to use it. We must take greater advantage of co-generation and combined heat and power opportunities. New electrical technologies of exceptional efficiency will be in high demand for the decades ahead. Storage systems linked to electricity systems will be increasingly important. Ontario has outstanding research capability and a vibrant manufacturing sector and could capture important future markets in energy technologies.

Policy goal for Ontario: *Research and development of electricity and related energy technologies.*

Energy efficient devices that can be turned down or off when the system is under stress (load control), and use of less electricity through changes in lifestyle, are an important part of the planning for future demand (see chapters 8 and 9). Yet energy use is determined in part by habit and culture, both of which are very resistant to change. Experience tells us that changing a technology choice is easier than changing a lifestyle or a business process. Research indicates that price and regulation are key tools for reducing electricity consumption in the long run (see chapters 12, 13, and 14). Incentives and information can provide a powerful boost to price-based energy conservation, but incentives and information are complements to, not substitutes for, price and regulatory tools. Improvements in efficiency that result from policies other than price increases can, in fact, lead to increased usage, called rebound, and even to an overall increase in consumption unless prices are increased concurrently.

Policy goals for Ontario: *Research on the social and behavioural science of energy consumption and conservation; and smart regulations to increase energy efficiency*

Wholesale Pricing

Today the wholesale price of electricity in Ontario is determined by adding a price set in the wholesale market, the regulated price of output from certain generation assets, a reflection of the cost of imports, and an allocation of a variety of other costs. From an economic point of view it would be desirable for the wholesale price to reflect, in each time period and at all locations in the system, the marginal social cost of generation of the highest-cost generator dispatched in that period, including the environmental harm associated with that generation. In theory, such pricing would encourage generators to invest where capacity is needed most and with the type of capacity that will best suit our needs (see chapter 10). However, market prices in Ontario have failed to stimulate investment in new generation, in part because of uncertainty about future government energy policy and in part because prices are too low. New generation has been built largely as a result of contracts with the Ontario Power Authority or of subsidies and above-market standard offer prices for renewable power. Higher wholesale prices reflecting marginal costs would at least encourage large consumers to moderate demand when generation costs are at their peak. At present, environmental costs are not included, and too much of the price is regulated. Consumer incentives would improve if we included environmental costs and if we increased the proportion of the power consumed by large customers that is subject to the market price from its currently low 20% (see chapter 10). Wholesale market design is a daunting challenge in part because there are many competing interests and no way to ensure that a revision would move closer to efficient pricing, rather than further away. Still, the potential efficiency gains from making wholesale prices fully reflect actual costs are substantial.

Policy goal for Ontario: *Modified wholesale market pricing that better reflects true costs including environmental costs in each time period.*

Consumer Pricing

Ontario has a long tradition of low-cost electricity based on low-cost generation from early hydroelectric projects. Today the cost of electricity generation, transmission, and distribution in Ontario is not particularly low compared to that of our U.S. neighbours, and it is clearly higher than in Quebec. In addition, the price paid by all consumers fails to incorporate all system costs or the cost of the environmental harm caused by some generation, such as the health effects of air pollution from coal-fired plants. If consumers and generators are to make sound decisions, we need to increase the price of electricity in Ontario. Moreover, the cost of supply varies by the season, the hour, and even the minute (when peak loads require that high-cost generation be brought on line). The simple flat price per kilowatt-hour that is paid by many electricity consumers under the Regulated Price Plan fails to confront those consumers with the true cost of their usage (see chapter 10). Studies show that consumers do respond to price, some more than others, and that some respond to information about price and conservation (see chapters 8 and 9). Yet consumer prices that vary by the hour are resisted, with the observation that demand is only modestly responsive to price and that consumers everywhere do not want to face electricity bills that fluctuate widely from month to month or to track prices by the hour and adjust usage accordingly. They may be willing to accept technology that allows their devices to respond to system prices automatically, such as turning down or off water heaters and air conditioners when system prices peak. Innovative pricing systems such as time-of-use pricing and critical peak pricing may be able to induce better responses.

Policy goals for Ontario: *Modified consumer pricing that better reflects true costs in each time period, without excessive bill volatility; and consumer access to technology that controls usage by specified appliances at times of system stress.*

The Political Challenge

Politics has never been far from electricity in Ontario. Electricity infrastructure is essential to our economic and physical well-being; it is enormous in economic and physical scale, and enormous in time scale, requiring decades-long planning. The electricity system is the largest machine in the province. Electricity policy in Ontario has a long history of being closely coupled to the province's economy and will have a long future being closely coupled to the province's environment (see chapter 18). From Red Lake to Yorkville, everybody uses electricity, and everybody's job and everybody's health depend on electricity. Electricity, at least a bit of it, has become as essential to life in Ontario as air, water, food, shelter, and sunshine.

Recently electricity policy has become more unstable. From 1995 to 1998 the government moved toward a competitive electricity market, but then it resisted selling parts of Ontario Hydro to create a competitive structure. When consumers protested the price increases, the government capped the price in November 2002. Private investors failed to secure government approval for the purchase or construction of generation stations. Soon the Ontario Power Authority was purchasing power because private investors had lost confidence in the market. The government has issued increasingly specific directions to the electricity sector, mandating the closure of coal plants and directing the purchase of electricity from specific types of generation. After the OPA had developed the Integrated Power System Plan, the Minister of Energy and Infrastructure in 2008 directed it to increase the renewable component of the plan. The 2009 *Green Energy and Green Economy Act* allows the Minister to direct the OPA to develop feed-in tariffs and to direct the price that will be paid. The act requires distributors and transmitters to provide connections to renewable and distributed generation. These are fundamental changes to the rules under which electricity planning has operated in Ontario, with more influence residing in the Minister and less in the Ontario Energy Board. Potential investors are uncertain about the way in which the rules of the Ontario electricity game are going to change in the future and are reluctant to risk their capital here. We have gone from promoting competition to discouraging it. This can only mean higher costs for electricity consumers.

Part of our social contract with the government includes the provision of safe, reliable, affordable electricity. Today we owe our thanks to those who have built and who run the electricity system from generation, through transmission and grid control, to distribution, and regulation of buildings and appliances; every day we have an ample, steady, and safe supply of affordable electricity. But, for the future, we need wise decisions to ensure that this remarkable record continues. We need to choose how best to make electricity that is 'safe' for the environment. We need to choose which forests and fields will be spoiled by new transmission lines or wind turbines. We need to agree to pay for *all* the costs, and that means paying more than we pay today. Everybody has an opinion and an interest (or a backyard, as in 'not in my backyard'). Is it any wonder that passions run high? But we must choose.

John Kenneth Galbraith, contradicting Bismarck, offered this wisdom: 'Politics is not the art of the possible. It consists of choosing between the disastrous and the unpalatable' (*Ambassador's Journal*, 1969). There are hard, sometimes unpalatable, choices that we, the people of Ontario, need to make to ensure the continued, vital supply of electricity. That is the political challenge.

The Mechanisms of Policy Formulation

Policymaking is intended to guide decision making; to marshal the facts; to bring orderly thinking to complex physical, social, political, and economic systems; and to create frameworks for deliberation, governance, regulation, or operation. Electricity policymaking in Ontario is complex. (Some have described it as confounded, intractable, even dysfunctional.)

Policymaking involves many stakeholders and many processes for consultation. A dizzying array of government authorities, ministries, electricity agencies, and electricity distribution agencies is involved in making policy for governing, regulating, and operating the electricity sector: the Ministry of Energy and Infrastructure, the Ministry of the Environment, Ontario Electricity Board, Ontario Power Authority, the Independent Electricity System Operator, Hydro One, Ontario Power Generation, Toronto Hydro, Barrie Hydro, and so on (see chapters 10 and 17). There are many large and millions of small consumers of electricity. There are numerous companies supplying goods and services to the electricity sector. There are numerous non-governmental organizations representing citizens' interests. Electricity policymaking does seem unwieldy.

The larger question concerns energy policymaking. If we are to use electricity to achieve a low-carbon economy, we must have policies and processes that give guidance on the best future for a wider range of energy-related issues, such as public transit, low-carbon automobiles, urban infrastructure, building heating and cooling, and freight movement. However, we must have policy mechanisms in which the political process sets broad objectives, and administrative bodies and stakeholders pursue those objectives in a predictable and stable manner. We need to avoid the risk of rapid-fire policy reversals in an industry where capital investments are expensive and will last forty years and more.

Policy goal for Ontario: *New mechanisms for stable economy-wide policymaking on energy.*

Communicating with the Public

There seems to be a shortfall in public understanding about the hard (sometimes unpalatable) choices that we face in the electricity/energy sector. Is it because the public are not interested, the issues are not of high enough concern, they do not think there is a problem, or they think it is not their problem? Or are there so many voices out there that they do not know who or what to trust? Is the public sufficiently informed – about shutting down coal-fired generation, the rising cost of generating electricity, the transmission constraints on renewables expansion in the north, or the need to put a price on carbon emissions? Forward-thinking policy can only gain public acceptance if it speaks to the needs of the people and does so in a clear and compelling way.

As Sean Conway puts it in chapter 18, 'it is absolutely critical that we find a mechanism for a better kind of public debate about the critical issues facing Ontario's electricity sector … engaging the public in a more sustained and honest discussion about the real choices we face and the costs and consequences of these choices.'

Policy goal for Ontario: *A new public conversation about energy.*

Box 19.1. A Summary of Proposed Policy Goals for Ontario

- Climate change guidance for all energy and energy-related policy
- Pricing of carbon emissions
- Low-carbon electricity
- Coherent encouragement of economical renewable generation
- Low-carbon electricity that provides energy for transportation and heating
- Electrically powered public transit for major urban centres
- Virtual elimination of pollutant emissions from coal-based electricity generating stations in Ontario
- Safe, secure, long-term storage of waste from nuclear power generation
- Careful evaluation and discussion of nuclear technology as to cost, environmental performance, economic performance, and the risks faced by the public, electricity consumers, and taxpayers
- Electricity infrastructure that provides a reliable and sufficient supply of electricity to meet demand
- Payment by electricity consumers of all the costs of electricity including environmental harm
- Research and development of electricity and related energy technologies
- Research on the social and behavioural science of energy consumption and conservation
- Smart regulations to increase energy efficiency
- Modified wholesale market pricing that better reflects true costs including environmental costs in each time period
- Modified consumer pricing that better reflects true costs in each time period without excessive bill volatility
- Consumer access to technology that controls usage by specified appliances at times of system stress
- New mechanisms for stable economy-wide policymaking on energy
- A new public conversation about energy
- New, independent institutions for energy policy research, analysis, and communication

Ontario Energy Policy Studies

There is a need for new, independent institutions for energy policy research, analysis, and communication. Such institutions should be based in universities and be highly multidisciplinary, including the disciplines of engineering, economics, law, political science, sociology, management, and environmental science. There should be a network of all Ontario universities working in the field. These new institutions should bring to the Ontario energy policy debate new ideas, new tools, and new people and engage thought leaders from a range of sectors: politicians and policymakers, academics and researchers, and experts foreign and domestic – from industry, utilities, not-for-profits, consultants, law, banking, neighbouring jurisdictions, and consumers big and small.

Policy goal for Ontario: *New, independent institutions for energy policy research, analysis, and communication.*

In Conclusion

Our intention in this chapter is to propose policy goals for the new conversation about energy. This is a critical time for concerted action to renew the mechanisms of policy debate and formulation on energy policy for Ontario.

Contributors

Steven Bernstein is an associate professor of political science, associate director of the Centre for International Studies, and director of the Masters Program in International Affairs, all at the University of Toronto. His research interests include global governance, global environmental politics, international political economy, globalization and internationalization of public policy, and international institutions. Publications include *The Compromise of Liberal Environmentalism* (Columbia University Press, 2001), *Global Liberalism and Political Order: Toward a New Grand Compromise?* (co-edited; SUNY Press, 2007), and *A Globally Integrated Climate Policy for Canada* (co-edited; University of Toronto Press, 2008); refereed articles in academic journals including *European Journal of International Relations*, *Journal of International Economic Law*, *Policy Sciences*, *Regulation and Governance*, *Canadian Journal of Political Science*, *Journal of International Law and International Relations*, and *Global Environmental Politics*; as well as a number of book chapters in edited volumes. His current research focuses on the problem of legitimacy in global governance.

Sean Conway is special advisor to the principal for external relations, and associate director of the Institute of Intergovernmental Relations in the School of Policy Studies, at Queen's University. He is also a policy advisor with the law firm Gowling Lafleur Henderson LLP and a public affairs analyst on TV Ontario. Mr Conway served in the Legislative Assembly of Ontario for twenty-eight years, from 1975 to 2003. He served in the Cabinet of Premier David Peterson (1985–90), holding the positions of minister of education, minister of colleges and universities, minister of mines, and government house leader. His most recent position in the Legislature was as Liberal party spokesperson for energy and electricity policy. Mr Conway holds a graduate degree in history from Queen's University and has taught at both Queen's and Wilfrid Laurier universities.

Donald Dewees is a professor of economics and a professor of law and is also appointed to the School of Public Policy and Governance at the University of Toronto. He served as director of research for the Ontario Royal Commission on Asbestos, as vice-dean and acting dean of the Faculty of Arts and Science of the University of Toronto, and as interim Chair of Economics. During 1998 he served as the vice-chair of the Ontario Market Design Committee, which advised the government on the introduction of competition into the electricity market in Ontario. His research and publications are in the fields of environmental economics, law and economics, and electricity restructuring. Dr Dewees has advised governments on environmental policy and electricity policy. His recent research has explored the use of emissions trading and emissions charges for pollution control, and issues of pollution control and pricing in electricity markets. He holds an engineering degree from Swarthmore College, a bachelor of laws from Harvard, and a doctorate in economics from Harvard.

Dominique Finon is senior research fellow in economics at the French National Center of Scientific Research (CNRS). From 1990 to 2002, he was head of the Institute of Energy Policy and Economics, and since 2003 he has been deputy director of the energy program at CNRS. His research focuses on energy modelling, economics and political economy of nuclear energy, research and development and innovation policies in the energy sector, policies for promotion of renewables and energy efficiency, and market reforms for electricity and gas industries. He has published numerous academic papers and several books. Dr Finon served as an expert for the European Commission, the government of France (the Ministry of Industry, the Ministry of Environment, the Energy Conservation Agency), French energy companies, and several multilateral development organizations.

Carolyn Fischer is a research fellow of Resources for the Future, an independent community of scholars conducting impartial research to enable policymakers to make sound choices, based in Washington, DC. Her research focuses on policy mechanisms and modelling tools that cut across environmental issues, including environmental policy design and technological change, international trade and environmental policies, and resource economics. In the areas of climate change and energy policy she has investigated the implications of different designs for emissions trading programs and has also conducted research on Corporate Average Fuel Economy standards, renewable portfolio standards, and energy efficiency programs. With Resources for the Future since 1997, Dr Fischer has taught at Johns Hopkins University and served as a staff economist for the Council of Economic Advisors. She holds a bachelor's degree in international relations and economics from the University of Pennsylvania and a doctorate in economics from the University of Michigan.

Mark Jaccard has been a professor in the School of Resource and Environmental Management at Simon Fraser University, Vancouver, since 1986 – interrupted from 1992 to 1997 while he served as chair and chief executive officer of the British Columbia Utilities Commission. Dr Jaccard holds a doctorate from the Energy Economics and Policy Institute at the University of Grenoble. Internationally, he is known for his work on the Intergovernmental Panel on Climate Change, the China Council for International Cooperation on Environment and Development, and the Global Energy Assessment (where he is the convening lead author for sustainable energy policy). Dr Jaccard is a member of Canada's National Roundtable on the Environment and the Economy and a research fellow at the C.D. Howe Institute.

Bryan Karney is a professor of civil engineering at the University of Toronto, where he has worked since 1987. He is also appointed to the School of Public Policy and Governance. He is currently associate dean for cross-disciplinary programs in the Faculty of Applied Science and Engineering. Dr Karney has spoken and written widely on subjects related to water, energy, environment, hydrology, climate change, engineering education, and ethics. He was an associate editor for the *Journal of Hydraulic Engineering* for the American Society of Civil Engineers from 1993 to 2005. He was selected Professor of the Year, in Civil Engineering, in 2000 and 2003 and won the Faculty of Applied Science and Engineering Teaching Award in 2002. He was one of the top ten finalists in TVO's Best Lecturer competition in 2007. Dr Karney has published three books, including the *Comprehensive Water Distribution Systems Analysis Handbook for Engineers and Planners* (2nd edition, MWH Press), as well as over two hundred papers and scientific contributions on topics ranging from water hammer, energy system performance, life-cycle analysis, hydrology, flows of frazil ice, and engineering education. He holds a bachelor of science degree in bio-resource engineering and a doctorate in civil engineering from the University of British Columbia and is a professional engineer.

Loren Lutzenhiser is a professor of urban studies and planning at Portland State University. His teaching interests include environmental policy and practice, energy behaviour and climate, technological change, urban environmental sustainability, and social research methods. Particular studies have considered variations across households in energy consumption practices, how energy-using goods are procured by government agencies, how commercial real estate markets work to develop both poorly performing and environmentally exceptional buildings, and how the 'greening' of business may be influenced by local sustainability movements and business actors. Dr Lutzenhiser recently completed a major study for the California Energy Commission, reporting on the behaviour of households, businesses, and governments in the aftermath of that state's 2001 electricity deregulation crisis, and he is currently exploring the relationships between household natural gas, electricity, gasoline, and water usage. He holds bachelor's and master's degrees in sociology from the University of Montana and a doctorate in sociology from the University of California, Davis.

Heather MacLean is an associate professor in the Department of Civil Engineering and the School of Public Policy and Governance at the University of Toronto. Her research focuses on energy systems analysis and, more specifically, life-cycle assessment of conventional and alternative fuels for the transportation sector, and low-carbon options for electricity generation. Her most recent work has evaluated the techno-economic and environmental aspects of bioenergy systems, including completing studies that examine the environmental implications of the use of agricultural residues and wood pellets for generating electricity in Ontario. Over the last decade she has worked closely with the automotive, electricity, and oil industries on bioenergy and oil sands research. In 2004 Professor MacLean was awarded an Early Researcher Award by the Government of Ontario for her work on evaluating alternative energy systems. She has also served on industry and government advisory committees. She holds a joint doctorate in civil engineering and engineering and public policy from Carnegie Mellon University.

James Meadowcroft holds the Canada Research Chair in Governance for Sustainable Development and is a professor of political science and public policy and administration at Carleton University. He has conducted research in the areas of political theory and environmental politics, specifically modern political ideologies, with a primary focus upon the evolution of contemporary ideological traditions. His research also examines the role of political science in environmental controversy, approaches to managing environmental dilemmas, and democracy and the environmental problematic. Professor Meadowcroft has written a series of works on sustainable development, focusing particularly on the challenge of planning for sustainable development, and on the implementation of plans, strategies, and initiatives in the industrialized countries since the 1992 Earth Summit in Rio de Janeiro. He is currently editor of *International Political Science Review*. He was educated at McGill University and received his doctorate from the University of Oxford.

Dean Mountain is a finance and business economics professor at McMaster University. Currently he is editor-in-chief of *Energy Studies Review*. He served as the director of the McMaster Institute for Energy Studies from 1998 to 2005 and as chair of the Finance and Business Economics Area of the DeGroote School of Business from 1997 to 2002. Dr Mountain specializes in energy economics, applied econometrics, and measurement of productivity and economies of scale in financial institutions. Current research interests include measuring energy conservation at the end-use level and the measurement of economies of scale in financial institutions. He has consulted in the area of energy economics, load research, statistical analysis of demand management programs, evaluation of new rate structures, the building and evaluation of aggregate and end-use energy-forecasting models, and measuring economies of scale in the financial sector. He holds a degree in economics and mathematics from McMaster University and a doctorate in economics from the University of Western Ontario.

Richard Newell is the Gendell Associate Professor of Energy and Environmental Economics at the Nicholas School of the Environment, Duke University. He is a research associate of the National Bureau of Economic Research and a university fellow of Resources for the Future. He has served as the senior economist for energy and environment on the President's Council of Economic Advisers, where he advised on policy issues ranging from automobile fuel economy and renewable fuels to management of the Strategic Petroleum Reserve. He has been a member of expert committees including the National Research Council committees on Energy Externalities, Energy Research and Development, Innovation Inducement Prizes, and Energy Efficiency Measurement Approaches. Dr Newell also served on the 2007 National Petroleum Council's Global Oil and Gas Study. He has served as an independent expert reviewer and advisor for many governmental, non-governmental, international, and private institutions including the OECD, Intergovernmental Panel on Climate Change, World Bank, National Commission on Energy Policy, U.S. Environmental Protection Agency, U.S. Department of Energy, U.S. Energy Information Administration, U.S. National Science Foundation. He received his doctorate from Harvard University.

Edward 'Ted' Parson is the Joseph L. Sax Collegiate Professor of Law and professor of natural resources and environment at the University of Michigan. His interests include environmental policy and its international dimensions, the political economy of regulation, the role of science and technology in policy and regulation, and the analysis of negotiations, collective decisions, and conflicts. His recent research has included projects on the relationship between environmental regulation and technological innovation, the policy implications of active carbon-cycle management, and methods of scientific and technical assessment for climate change. Dr Parson has served on multiple senior advisory bodies in Canada and the United States, including the panel of the National Academy of Sciences' project on 'America's Climate Choices.' He holds degrees in physics from the University of Toronto and in management science from the University of British Columbia, and a doctorate in public policy from Harvard.

Doug Reeve was the co-chair of the conference 'Current Affairs: Perspectives on Electricity Policy for Ontario,' held in Toronto on 4–5 June 2007, on which this book is based. He is professor and chair in the Department of Chemical Engineering and Applied Chemistry at the University of Toronto. Dr Reeve was the founding director of the University of Toronto Pulp and Paper Centre, serving from 1987 until 2001. In 2002 he established a student leadership development program in the department and is presently co-leader of Leaders of Tomorrow, a leadership development program that reaches across the Faculty of Applied Science and Engineering. As chair of the Task Force on Engineering and Public Policy (EPP) he has been engaged in the development of concepts for teaching EPP at the undergraduate and post-graduate level. He has been appointed to the new University of Toronto School of Public Policy and Governance. Professor Reeve is a member of the university's Governing Council. He has received numerous awards for his contributions; most recently, he received the Carolyn Touhy Impact on Public Policy Award from the university, and in 2006 he was named Fellow of the Canadian Academy of Engineering. He received his doctorate in chemical engineering from the University of Toronto and is a professional engineer.

Fereidoon Sioshansi is the president of Menlo Energy Economics, a consulting firm based in San Francisco, California, and serving the energy sector. Dr Sioshansi's professional experience includes working at Southern California Edison Company, the Electric Power Research Institute, National Economic Research Associates, and most recently, Global Energy Decisions, now called Ventyx. His recent edited books include *Electricity Market Reform: An International Perspective* (2006), *Competitive Electricity Markets: Design, Implementation, Performance* (2008), and Generating Electricity in a Carbon Constrained World (2009). A fourth volume, *Smart Living with a Low Ecological Footprint*, is forthcoming in 2010. Dr Sioshansi is the editor and publisher of *EEnergy Informer*, a monthly newsletter with wide international circulation. He is on the editorial advisory board of *The Electricity Journal*, where he writes the 'Electricity Currents' section. A frequent contributor to *Energy Policy*, he serves on the editorial board of *Utilities Policy*. He has degrees in engineering and economics, including a master of science degree and a doctorate in economics from Purdue University.

Steven Sorrell is a senior fellow with the SPRU Energy Group (Science and Technology Policy Research) at the University of Sussex in the United Kingdom. He trained as an electrical engineer and spent four years working in industrial research and development laboratories before gaining a master's degree in science and technology policy in 1990. Since joining SPRU, he has undertaken a range of research on energy and environmental policy, with particular focus on energy modelling, energy efficiency, and emissions trading. Mr Sorrell has published three books, eighteen papers in refereed journals, twelve book chapters, and more than one hundred research reports. He has acted as consultant to the European Commission, UK government departments, the UK Environment Agency, the UK Sustainable Development Commission, and private sector and non-governmental organizations. He is currently deputy director of the Sussex Energy Group (SEG) and is co-managing the technology and policy assessment function of the UK Energy Research Centre (UKERC). Mr Sorrell's current research projects include energy use and sustainability in UK freight transport (SEG), trials of smart metering technology (UK Department of Energy and Climate Change and EDF), and global oil depletion (UKERC).

Ulrik Stridbaek has worked as a chief economist, regulatory affairs, Group R&D, at DONG Energy since August 2008. Regulatory framework, energy policy, and business development in Denmark and Europe are the main areas of his attention. He worked as a senior policy advisor, electricity markets, at the International Energy Agency (IEA) for the previous four years. At the IEA he focused on energy sector reform, market design, regulation, investment, demand response, and trends in electricity sector technologies. Prior to the IEA, Mr Stridbaek worked as an economist for seven years at the Danish Transmission System Operator, Eltra. He has a master of science degree in economics from the University of Aarhus, Denmark, and the Universidad de Barcelona, Spain.

Michael Trebilcock is University Professor and chair in law and economics at the University of Toronto. He joined the Faculty of Law at the University of Toronto in 1972. Professor Trebilcock specializes in law and economics, international trade, and contract and commercial law. He has served as national vice-president of the Consumers' Association of Canada, chair of the Consumer Research Council, and research director of the Professional Organizations' Committee for the Government of Ontario. From 1982 to 1986 he was a member of the Research Council of the Canadian Institute of Advanced Research. Professor Trebilcock was awarded the Owen Prize in 1989 by the Foundation for Legal Research for his book *The Common Law of Restraint of Trade*. During 1998 he served as the research director of the Ontario (Electricity) Market Design Committee. He received his bachelor of laws degree from New Zealand in 1961, and his master of laws degree from Adelaide in 1962; he was called to the Bar of New Zealand in 1964 and the Bar of Ontario in 1975.

George Vegh is the head of McCarthy Tétrault's Toronto energy regulatory practice. Prior to joining the firm, he was general counsel of the

Ontario Energy Board. Mr Vegh teaches energy law and policy at the University of Toronto Law School as an adjunct professor and as an associate member of the graduate faculty. He is also a member of the board of directors of the Ontario Energy Association. Mr Vegh received his bachelor of laws degree from the University of Toronto in 1989, and his master of laws degree from Osgoode Hall Law School in 1996. He was called to the Ontario Bar in 1991.

Workshop Contributors

Tom Adams is an independent energy and environmental advisor. He has held a variety of senior responsibilities including executive director of the consumer and environmental think tank Energy Probe from 1996 until September 2007, membership on the Ontario Independent Electricity Market Operator's board of directors, and membership on the Ontario Centre for Excellence for Energy's board of management. He has lectured in energy studies at the University of Toronto, and his guest columns have appeared in most major Canadian newspapers. He currently sits on the board of directors of the agricultural research and development organization REAP Canada. Mr Adams has presented expert testimony before many regulatory tribunals in Canada on a wide variety of energy subjects. He has made presentations to legislative committees in Ontario and New Brunswick; academic, regulatory, and trade conferences; the Atomic Energy Control Board; and the Canadian Nuclear Safety Commission. His profile appears in *Canadian Who's Who.*

Branko Terzic is global and U.S. regulatory policy leader at Energy & Resources Deloitte Services LP. He serves as chairman of the United Nations ECE Ad Hoc Group of Experts on Cleaner Electricity Production from Coal and Other Fossil Fuels; the advisory council of the North American Energy Standards Board; National Petroleum Council; Executive Council Energy Efficiency Forum; and Bordeaux Energy Colloquium. Mr Terzic is a fellow of Royal Society Arts, former commissioner of the Federal Energy Regulatory Commission, commissioner of Wisconsin Public Service Commission, and chief executive officer and president of Yankee Energy System. He wrote the energy chapter in *The World Crisis: The Way Ahead after Iraq* (edited by Robert Harvey and Geoffrey Howe, 2008). He holds a bachelor of science degree in Engineering from the University of Wisconsin- Milwaukee and is a professional engineer.

Acknowledgments

Current Affairs grew out of a new vision of ways in which to discuss and develop electricity policy in Ontario. Many people were involved in framing the questions and in developing and organizing the workshop that was the basis of this book, and in developing the book itself. In particular we wish to thank Steven Bernstein, Bashir Bhana, Paul Burke, Phil Byer, Nicolle Geneau, Hans Hesse, Heather MacLean, Julia McNally, Jonathan Norman, Michael Reid, Mark Stabile, Bob Stasko, and Jessica Woolard. In addition, Mark Stabile, through the new forum of the School of Public Policy and Governance, showed us a new way of organizing and leading a large and diverse team to tackle a complex policy regime. We thank Zora Anaya, Petra Jory, Joan Rogers, and Cindy Tam for their administrative assistance.

We are especially grateful to the speakers and discussants who contributed to the workshop and to this volume for sharing their vast and varied experience and their understanding with us.

This project received financial support from Hydro One, the Independent Electricity System Operator, the Ontario Ministry of Energy, Ontario Power Authority, Ontario Power Generation, and the Ontario Centre of Excellence for Energy. They allowed us complete freedom in choosing topics and speakers and are not responsible for any of the views expressed in the book.

We are grateful to the University of Toronto and the following divisions of the university: the School of Public Policy and Governance, the Faculty of Applied Science and Engineering, the Department of Economics, and the Faculty of Law.